Troubleshooting Process
Plant Control

Troubleshooting Process Plant Control
Other Books by Norman P. Lieberman

- *Troubleshooting Refinery Operations*—Penn Well Publications
- *Troubleshooting Process Operations 4th Edition*—PennWell Publications
- *A Working Guide to Process Equipment* (with E. T. Lieberman)—3rd Edition—McGraw Hill Publications
- *Troubleshooting Natural Gas Processing* (order by e-mail at norm@lieberman-eng.com)
- *Process Design for Reliable Operations 3rd Edition* (order by e-mail at norm@lieberman-eng.com)

Copies of the first three texts are best ordered from the publishers, but may be ordered through us. E-mail (norm@lieberman-eng.com). *Troubleshooting Refinery Operations* (1980) has been incorporated into *Troubleshooting Process Operations* and *Troubleshooting Natural Gas Processing*.

Troubleshooting Process Plant Control

Norman P. Lieberman

A John Wiley & Sons, Inc., Publication

DISCLAIMER

Company names and the names of individuals used in this book are entirely fictitious. I have selected company names and the names of colleagues entirely at random. Technically, the stories I have related are correct but are sometimes combinations of actual incidents. Any similarity to actually existing events, process plants, or individuals is purely a coincidence. Also, I sometimes have related projects executed by operators or other engineers as if I had myself originated, rather than just participated in, such events. I have consistently failed throughout this text to assign credit to other individuals for their ideas, which I have not stolen but only borrowed. Any names of actual individuals or process plants or refineries have been chosen at random and do not correspond to any real events connected with the people or locations mentioned in my text.

Norman P. Lieberman

Published by John Wiley & Sons, Inc., Hoboken, New Jersey
Published simultaneously in Canada

For general information on our other products and services or for technical support, please contact our Customer Care Department within the United States at (800) 762-2974, outside the United States at (317) 572-3993 or fax (317) 572-4002.

Wiley also publishes its books in a variety of electronic formats. Some content that appears in print may not be available in electronic formats. For more information about Wiley products, visit our web site at www.wiley.com.

Library of Congress Cataloging-in-Publication Data:
Lieberman, Norman P.
   Troubleshooting process plant control / Norman P. Lieberman.
      p.   cm.
   ISBN 978-0-470-42514-5 (cloth)
1. Petroleum refineries–Maintenance and repair.   I. Title.
   TP690.3.L534 2009
   665.5028′8–dc22

2008032181

10 9 8 7 6 5

Dedication

One of life's little pleasures is working with quality and dedicated people, such as April Montecino Winn and Phil Negri. Persistently and consistently they have worked with my wonderful draftsman, Roy Williams, to bring order to my scribbled manuscript and illegible drawings to produce this book. Inspired by the Creator, these three wonderful people have brought Order out of Chaos.

I, too, have been inspired by my equally wonderful partner in life, Liz. She is a light unto my life.

Contents

PREFACE WARNING:
The Surgeon General Has Determined That This Book Is a Fraud

Dr. R. K. Sudkamp, Ph.D., having reviewed this text, reports that this book's claim to be a technical work on the exalted subject of Process Control has no basis. Further, findings by the Federal Communications Commission have determined that this book cannot claim to be a Process Control text as enumerated below:

- The entire book is free of complex mathematics. Even simple equations are rarely encountered.
- The text is much too easy to read for any respectable technical book.
- Much of the so-called "advanced technology" described by the author is 40 years old and is already in widespread practice in the process industry.

While Professor Sudkamp notes the potential usefulness of this book to solve practical plant process problems, he also observes this is more than offset by its total lack of applicability as a postgraduate university text. Professor Sudkamp, of the University of Stockholm, has reported a deep loss of personal dignity as a consequence of his exposure to Mr. Lieberman's book, which he considers to be an insult to his lofty intellect.

Introduction—
A History of Positive
Feedback Loops

Process Control Engineering is the most important branch of Chemical Engineering. Ask any panel board operator in a petrochemical plant or refinery. The P&IDs (Process and Instrumentation Diagrams) are the definitive engineering documents describing how a plant works. The Process Control Engineer has the ultimate responsibility for creating, maintaining, and interpreting the P&IDs. It is his job—and perhaps his most important function—to explain to the panel operator how the control valves interact with the process plant to achieve unit stability.

I don't know why I wrote "his job," because half of the Process Control Engineers I work with are women. Women often make better control engineers and panel board operators than men, because they are more patient. Men often are driven to reach some distant goal quickly. Women, being patient, will take a more measured approach to restore stability during a process upset.

For example, one question I am frequently asked by younger male process engineers and console operators is how to meet girls. One fellow, Jake, described his problem. "I met this lady at a bar, Norm. I introduced myself and asked her name."

"Hi, I'm Linda."

"Can I buy you a drink, Linda?"

"Actually, Jake, I'm perfectly capable of buying my own drink."

"Norm, I hate when girls blow me off like that. I couldn't think what else to say."

"Okay, Jake. Here's the correct line," I advised. "Say Linda, I'm conducting a survey. Which is better, negative feedback or positive feedback?"

"Well, Jake," Linda will answer, "I'm a positive-thinking person. Therefore I'm sure that positive feedback is best."

"Not so, Linda," you'll say, "As a highly paid Process Control Engineer I've found negative feedback is best. Do you mind if I explain?"

"Okay, Jake," she'll say, "Maybe I will have a small scotch and soda."

"Let's say, Linda, you're driving your car uphill. The car is in automatic cruise control. As the gradient increases, the car slows. The gas pedal is automatically depressed to accelerate the car. But as the car engine slows, the amount of combustion air drawn into the engine is reduced. As the air flow drops, the incremental gasoline injected into the cylinders does not burn. But as the gasoline vaporizes, it also cools the engine's cylinders. This reduces the cylinder pressure and hence the force acting on the pistons heads. This decreases the engine horsepower. The engine and car both slow further. This signals the automatic cruise control to inject more gasoline into the cylinders. As the engine is already limited by combustion air and not by fuel, the extra gasoline just makes the problem worse. If you don't switch off the cruise control and return to manual operation, the car will stall out and the engine will flood with gasoline.

"The problem, Linda, is positive feedback. The loss of engine speed reduced the air flow to the engine and also automatically caused more gasoline to flow into the engine. The problem fed upon itself. That's why we call this a positive feedback loop. What the Process Control Engineer wants is negative feedback. For a negative feedback loop to work, the engine cannot be limited by the combustion air flow."

"Linda, I run into this problem all the time. It happens in distillation towers, vacuum jet systems, and fired heaters. You see, control loops for process equipment only function properly when they are running in the range of a negative feedback response. Positive feedback is dangerous in that it leads to process instability."

Jake has subsequently tried this approach with Gloria, Janet, and Carol, all with the same result. When he comes to the part about process instability, each lady suddenly remembers an important appointment and rushed out of the bar.

NOTE TO READERS

I've written this book for three groups of people:

- The experienced plant panel board operator
- The Process Control Engineer who, with his degree in hand, must now face up to the real world

- The Process Engineer who must design control loops for new or revamped units

The text is very basic and very simple. Lots of new control concepts are presented. But they are all based on old technical concepts. Likely, you are reading this text because you are one of the 12,000 students who have already attended one of my seminars, or read one of my other books, or viewed my videotapes on process technology.

If not, let me tell you something about myself. My descriptions about control loops, process control optimization, process instrumentation, and control valves in this book are based on my personal experience which encompasses 44 years. Except for the story about Linda and Jake. I just made that up.

LATER THAT EVENING

My wife and partner, Liz, has just read the story about Linda and Jake. Liz says, "Can't you give an example of a positive feedback loop without all of the sexist stupidity?" Okay, I will:

- CO_2 accumulates in the atmosphere.
- The rate of CO_2 accumulation between 1968 and 2008 was 0.51% compounded annually.
- Global warming has increased by 1 °F since the early 1900s, including the surface of the oceans.
- For the 1960–2000 period, sulfur emissions from oil and coal combustion generated atmospheric sulfates, which reflected sunlight and suppressed global warming.
- Sulfates are scrubbed from the atmosphere by rain every year. Sulfur emissions in the past decade have mostly been stopped, and the rate of global warming has increased.
- Only 60% of the CO_2 generated from combustion of oil, gas, coal, and cement production has been accumulating in the atmosphere. The rest is absorbed in ocean surface waters.
- The ocean surface water is becoming more acidic and hotter. Both factors reduce the solubility of CO_2 in water.
- As the land becomes warmer, ice and snow melt in Greenland and the Antarctic. The Earth becomes less reflective to sunlight.
- As the Earth becomes warmer, methane emissions from frozen tundra, peat bogs, and offshore hydrate deposits increase. Methane per mole is 23 times more powerful a greenhouse gas than CO_2.

- In the next few decades the ocean surface waters will become an emitter rather than an absorber of CO_2 because of warming and increased acidity.
- Accumulating CO_2, has in the absence of increased sulfate concentration in the atmosphere, accelerated global warming.
- Accelerating global warming will reduce the absorption capacity of the ocean surface waters for CO_2 and increase methane emissions.
- Warmer global temperature will increase humidity. Water is a bigger greenhouse gas than either CO_2 or methane. The humidity effect will build upon itself.[1]

Liz, that's a real positive feedback loop. But I don't think Jake will get very far in finding a new girlfriend with this grim tale of global warming. Especially if Jake tells Linda the end of the story.

Eventually the small percentage of ocean surface waters that have warmed and become acidic will mix with deeper, cooler, neutral pH layers of the ocean, which contain the vast bulk of the planet's water. This will stop the positive feedback loop and global warming. So my story does have a happy ending—if Linda and Jake can just wait a few thousand years.

REFERENCE

1 Desonie, Dana, "Climate—Our Fragile Planet," Chelsea House Publishers, 2006.

1

Learning from Experience

An old Jewish philosopher once said, "Ask me any question, and if I know the answer, I will answer it. And, if I don't know the answer, I'll answer it anyway." Me too. I think I know the answer to all control questions. The only problem is, a lot of my answers are wrong,

I've learned to differentiate between wrong and right answers by trial and error. If the panel board operator persistently prefers to run a new control loop I've designed in manual, if he switches out of auto whenever the flow becomes erratic, then I've designed a control strategy that's wrong. So, that's how I've learned to discriminate between a control loop that works and a control strategy best forgotten.

Here's something else I've learned. Direct from Dr. Shinsky, the world's expert on process control:

- "Lieberman, if it won't work in manual, it won't work in auto."
- "Most control problems are really process problems."

I've no formal training in process control and instrumentation. All I know is what Dr. Shinsky told me. And 44 years of experience in process plants has taught me that's all I need to know.

Troubleshooting Process Plant Control, by Norman P. Lieberman
Copyright © 2009 John Wiley & Sons, Inc.

LEARNING FROM PLANT OPERATORS

My first assignment as a Process Engineer was on No. 12 Pipe Still in Whiting, Indiana. This was a crude distillation unit. My objective was to maximize production of gas oil, as shown in Figure 1-1. The gas oil had a product spec of not more than 500 ppm asphaltines. The lab required half a day to report sample results. However, every hour or two the outside operator brought in a bottle of gas oil for the panel board operator. The panel operator would adjust the wash oil flow, based on the color of the gas oil.

While plant supervision monitored the lab asphaltine sample results, plant operators ignored this analysis. They adjusted the wash oil rate to obtain a clean-looking product. The operators consistently produced a gas oil product with 50–200 ppm asphaltines. They were using too much wash oil. And the more wash oil used, the lower the gas oil production.

I mixed a few drops of crude tower bottoms in the gas oil to obtain a bottle of 500 ppm asphaltine material. I then instructed the panel board operators as follows:

- If the sample from the field is darker than my standard bottle, increase the wash oil valve position by 5%.
- If the sample of gas oil from the field is lighter than my standard, decrease the wash oil valve position by 3%.
- Repeat the above every 30 minutes.

The color of gas oil from a crude distillation unit correlates nicely with asphaltine content. The gas oil, when free of entrained asphaltines, is a pale yellow. So it seems that my procedure should have worked. But it didn't. The operators persisted in drawing the sample every one to two hours.

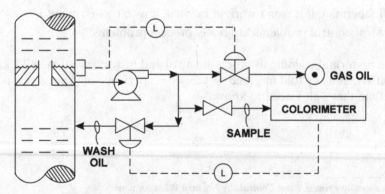

Figure 1-1 Adjusting wash oil based on gas oil color

So, I purchased an online colorimeter. The online colorimeter checked whether the gas oil color was above or below my set point. With an interval of 10 minutes it would move the wash oil valve position by 1%. This never achieved the desired color, but the gas oil product was mixed in a tank. The main result was that gas oil production was maximized consistent with the 500 ppm asphaltine specification.

One might say that all I did was automate what the operators were already doing manually, that all I accomplished was marginally improving an existing control strategy by automating the strategy. But in 1965 I was very proud of my accomplishments. I had proved, as Dr. Shinsky said, "If it does work on manual, we can automate it."

LEARNING FROM FIELD OBSERVATIONS

Forty years ago I redesigned the polypropylene plant in El Dorado, Arkansas. I had never paid much attention to control valves. I had never really observed how they operate. But I had my opportunity to do so when the polypropylene plant was restarted.

The problem was that the purchased propylene feed valve was too large for normal service. I had designed this flow for a maximum of 1600 BSD, but the current flow was only 100 BSD. Control valve response is quite nonlinear. Nonlinear means that if the valve is open by 5%, you might get 20% of the flow. If you open the valve from 80% to 100%, the flow goes up by an additional 2%. Nonlinear response also means that you cannot precisely control a flow if the valve is mostly closed. With the flow only 20% of the design flow, the purchased propylene feed was erratic. This resulted in erratic reactor temperature and erratic viscosity of the polypropylene product.

The plant start-up had proceeded slowly. It was past midnight. The evening was hot, humid, and very dark. I went out to look at the propylene feed control valves. Most of the flow was coming from the refinery's own propylene supply. This valve was half open. But the purchased propylene feed valve was barely open. The valve position indicator, as best I could see with my flashlight, was bumping up and down against the "C" (closed) on the valve stem indicator.

The purchased propylene charge pump had a spillback line, as shown in Figure 1-2. I opened the spillback valve. The pump discharge pressure dropped, and the propylene feed valve opened to 30%. The control valve was now operating in its linear range.

Now, when I design a control valve to handle a large reduction in flow, I include an automated spillback valve from pump discharge to suction. The spillback controls the pump discharge pressure to keep the FRC valve between 20% and 80% open. Whenever I sketch this control loop I recall that dark night in El Dorado. I also recall the value of learning even the most basic control principles by personal field observations.

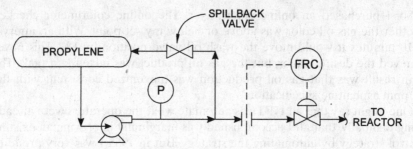

Figure 1-2 *Opening spillback to keep FRC valve in it's linear operating range*

LEARNING FROM MISTAKES

Adolf Hitler did not always learn from his mistakes. For example, he once ordered a submarine to attack the Esso Lago Refinery in Aruba. The sub surfaced in the island's harbor and fired at the refinery. But the crew neglected to remove the sea cap on the gun's muzzle. The gun exploded and killed the crew.

I too had my problems in this refinery. The refinery flare was often very large and always erratic. The gas being burned in the flare was plant fuel. The plant fuel was primarily cracked gas from the delayed coker, supplemented (as shown in Fig. 1-3) by vaporized LPG. So much fuel gas was lost by flaring that 90% of the Aruba's LPG production had to be diverted to fuel, via the propane vaporizer.

I analyzed the problem based on the dynamics of the system. I modeled the refinery's fuel consumption vs. cracked gas production as a function of time. The key problem, based on my computer system dynamic analysis, was the cyclic production of cracked gas from the delayed coker complex. My report to Mr. English, the General Director of the Aruba Refinery, concluded:

1. The LPG vaporizer was responding too slowly to changes in cracked gas production from the delayed coker.
2. The natural log of the system time constants of the coker and vaporizer were out of synchronization.
3. A feed-forward, advanced computer control based on real-time dynamics would have to be developed to bring the coker vaporizer systems into dynamic real-time equilibrium.
4. A team of outside consultants, experts in this technology, should be contracted to provide this computer technology.

Six months passed. The complex, feed-forward computer system was integrated into the LPG makeup and flaring controls shown in Figure 1-3. Adolf

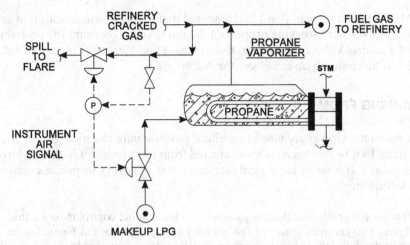

Figure 1-3 Unintentional flaring caused by malfunction of LPG makeup control valve is an example of split-range pressure control

Hitler would have been more sympathetic than Mr. English. The refinery's flaring continued just as before. Now what?

Distressed, discouraged, and dismayed, I went out to look at the vaporizer. I looked at the vaporizer for many hours. After a while I noticed that the fuel gas system pressure was dropping. This happened every three hours and was caused by the cyclic operation of the delayed coker. This was normal.

The falling fuel gas pressure caused the instrument air signal to the LPG makeup valve to increase. This was an "Air-to-Open" valve (see Chapter 11), and more air pressure was needed to open the propane flow control valve. This was normal.

BUT, the valve position itself did not move. The valve was stuck in a closed position. This was not normal.

You will understand that the operator in the control room was seeing the LPG propane makeup valve opening as the fuel gas pressure dropped. But the panel board operator was not really seeing the valve position; he was only seeing the instrument air signal to the valve.

Suddenly, the valve jerked open. The propane whistled through the valve. The local level indication in the vaporizer surged up, as did the fuel gas pressure. The flare valve opened to relieve the excess plant fuel gas pressure and remained open until the vaporizer liquid level sank back down, which took well over an hour. This all reminded me of the sticky side door to my garage in New Orleans.

I sprayed the control valve stem with WD-40, stroked the valve up and down with air pressure a dozen times, and cleaned the stem until it glistened. The next time the delayed coker cycled, the flow of LPG slowly increased to catch the falling fuel gas pressure, but without overshooting the pressure set point and initiating flaring.

My mistake had been that I had assumed that the field instrumentation and control valves were working properly. I did not take into account the probability of a control valve malfunction. But at least I had learned from my mistake, which is more than you could say for Adolf Hitler.

LEARNING FROM THEORY

Northwestern University has an excellent postgraduate chemical engineering program. I know this because I was ejected from their faculty. I had been hired to present a course to their graduate engineers majoring in process control My lecture began:

"Ladies and gentlemen, the thing you need to know about control theory is that if you try to get some place too fast, it's hard to stop. Let's look at Figure 1-4. In particular, let's talk about tuning the reflux drum level control valve.

Do I want to keep the level in the drum close to 50%, or doesn't it matter? As long as the level doesn't get high enough to entrain light naphtha into fuel gas, that's okay. What is not okay is to have an erratic flow feeding the light naphtha debutanizer tower.

On the other hand, if the overhead product was flowing into a large feed surge drum, than precise level control of the reflux drum is acceptable.

In order for the instrument technician to tune the level control valve, you have to show him what you want. To do this, put the level valve on manual. Next, manipulate the light naphtha flow to permit the level swings in the reflux drum you are willing to tolerate. But you will find that there is a problem. If you try to get back to the 50% level set point quickly you will badly overshoot your level target.

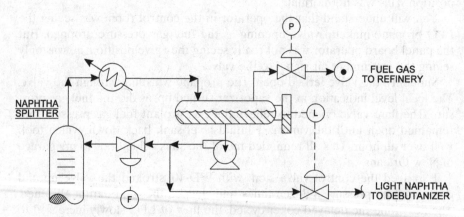

Figure 1-4 Tuning a level control valve depends on what is downstream

If you return slowly to the set point, it's easy to reestablish the 50% level target. However, the level will be off the target for a long time.

In conclusion, ladies and gentlemen, tuning a control loop is a compromise between the speed at which we wish to return to the set point and our tolerance to overshooting the target. To establish the correct tuning criteria, the control loop is best run on manual for a few hours by the Process Control Engineer. Thank you. Class adjourned for today."

My students unfortunately adjourned to Dean Gold's office. Dean Gold lectured me about the student's complaints.

"Mr. Lieberman, did you think you were teaching a junior high school science class or a postgraduate course in process control?"

And I said, "Oh! Is there a difference?"
So that's how I came to be ejected from the faculty of Northwestern University after my first day of teaching.

LEARNING FROM RELATIONSHIPS

My ex-girlfriend used to tell me, "Norm, the reason we get along so well is that I give you a lot of positive feedback." From this I developed the impression that positive feedback is good. Which is true in a relationship with your girlfriend. But when involved in a relationship with a control loop, we want negative feedback. Control logic fails when in the positive feedback mode of control. For example:

- **Distillation**—As process engineers and operators we have the expectation that reflux improves fractionation, which is true, up to a point. That point where more reflux hurts fractionation instead of helps is called the "incipient flood point." Beyond this point, the distillation tower is operating in a positive feedback mode of process control. That means that tray flooding reduces tray fractionation efficiency. More reflux simply makes the flooding worse.
- **Fired Heaters**—Increasing furnace fuel should increase the heater outlet temperature. But if the heat release is limited by combustion air, then increasing the fuel gas will reduce the heater outlet temperature. But as the heater outlet temperature drops, the automatic control calls for more fuel gas, which does not burn. As the heater outlet temperature continues to fall, because combustion is limited by air, the outlet temperature drops further. The heater automatic temperature control loop is now in the positive feedback mode of control. As long as this control loop is on auto, the problem will feed upon itself.

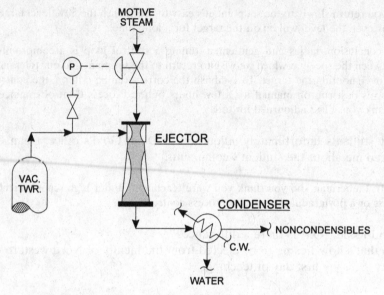

Figure 1-5   *Too much steam flow causes a loss in vacuum*

- **Vacuum Ejector**—Some refineries control vacuum tower pressure by controlling the motive steam flow to the steam ejector. As the steam pressure
  and flow to the ejector increases, the ejector pulls a better vacuum, as
  shown in Figure 1-5, but as the steam flow increases, so does that load on
  the downstream condenser. As the condenser becomes overloaded, the
  ejector discharge pressure rises. At some point the increased discharge
  pressure adversely affects the ejector's suction pressure. A further increase
  in motive steam will make the vacuum worse, instead of better. As the
  vacuum gets worse, the control loop calls for more steam. Having now
  entered the positive feedback mode of control, the problem feeds upon
  itself.

Many control loops are subject to slipping into a positive feedback loop.
The only way out of this trap is to switch the controls to manual and slowly
climb back out of the trap. Once you guess (but there is no way to know for
sure) that you are in the safe, negative feedback mode of control, you can then
safely switch back to automatic control.

2

Process Control Parameter Measurement

I mentioned in Chapter 1 that I was ejected from the faculty of Northwestern University after teaching a single class. This was not the end of my academic career. I was also an instructor at Louisiana State University. Dr. Dillard Smythe had hired me on a trial basis to conduct a process control course for undergraduate chemical engineers. My course was excellent, but judge for yourself.

"Ladies and gentlemen. Welcome to Process Control 101. The course is divided into two segments:

- **Segment One**—Measuring Process Control Parameters
- **Segment Two**—Designing Control Loops for Process Parameters

We must measure the parameter before we can control the parameter. That's why we will study measurement first.

The Nazi army was able to initially defeat the allied armies in World War II because of the superior use of tanks. It wasn't that the German tanks were better than the Allied tanks. It was that the Germans had excellent FM radios in their tanks. The data supplied from forward units enabled senior commanders to coordinate the Panzer attack. That is, the limiting factor for any control strategy is the quality of the data. Garbage in; garbage out.

I plan to discuss measurement techniques and problems for the following process parameters:

Troubleshooting Process Plant Control, by Norman P. Lieberman
Copyright © 2009 John Wiley & Sons, Inc.

- Liquid levels
- Temperature
- Pressure
- Differential pressure
- Flow

My experience is limited to continuous processes, but excludes solids and high-viscosity fluids. So let's limit our study accordingly. My experience in the petrochemical and refining industry has taught me that most control problems are a consequence of improper parameter measurement, most especially levels.

HOW ARE LIQUID LEVELS MEASURED?

Most liquid level measurement is made by a level-trol. The level-trol is served by two pressure transducers. A pressure transducer is a mechanical device that converts a pressure in an electronic signal. Car engines have a transducer to measure the engine oil pressure.

Figure 2-1 shows the arrangement of the pressure transducers, one at the top and one at the bottom of the level-trol. The level-trol is a pipe a few feet long. The difference in the electrical output between the dual pressure trans-ducers is proportional to the difference in pressure between the top and bottom of the level-trol. This delta P is caused by the head of liquid in the level-trol. The electrical output generated by this pressure difference is called

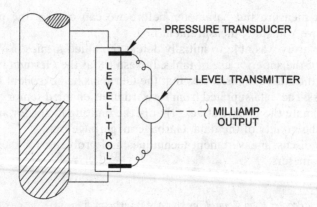

Figure 2-1 *Measuring levels by sensing liquid head pressure*

the "milliamp output of the level indicator." The level indication is really a measure of the head pressure in the level-trol pipe. Head pressure, DP, is calculated as DP = (Height) • (Density).

The level-trol cannot discriminate between height (i.e., liquid level) and density (i.e., specific gravity). The Process Control Engineer has to specify the liquid's density or specific gravity (SG). Let's say the specified SG = 0.80 and the calculated level from the delta P output from the level-trol is 45%. This 45% level is displayed on the panel in the control room.

The 45% level multiplied by the specific gravity of 0.80 SG results in a delta P of 36 units of differential pressure:

$$(45\%) \cdot (0.80) = 36 \text{ units of delta P}$$

But now, a new situation has developed. The feed to the vessel has become lighter. Or the bottom's product has become hotter. Or the liquid in the vessel is aerated. For some reason, the specific gravity has dropped from 0.80 SG to 0.60 SG.

Assume that the delta P output from the level-trol is constant at 36 units of differential pressure. Thus the indicated level in the control room is still 45%. But the real level has increased to 60%. That is, the 60% level multiplied by the specific gravity of 0.60 SG results in a delta P of 36 units of differential pressure:

$$(60\%) \cdot (0.60) = 36 \text{ units of delta P}$$

Thus the level in the vessel has gone up by one-third, but the panel level indication has not changed. A reduction in fluid density will therefore result in an automatic increase in the level in a vessel as long as the level control loop is in automatic. This precise problem has resulted in explosions and fires; death and disaster throughout the process industry.

One way that we deal with this problem in refineries is with radiation level detection, which is expensive, complex, and potentially dangerous because of problems with handling radioactive materials. We could also mathematically correct the indicated level for changes in density by a closed-loop computer control. But this can only be done if we know the new fluid density. In cases where the density has dropped because of aeration, which is a common problem, we do not know the aerated density in the bottom of the vessel.

So, in conclusion, what is the answer? The answer is—there is no answer! But it is certainly something for the Process Control Engineer to worry about. Fifteen people were killed at the BP Refinery in Texas City because no one understood this problem in a naphtha fractionation tower, which erupted gasoline from its relief valve.

HOW ARE TEMPERATURES MEASURED?

I always keep a spare thermocouple at home. I need it in case the pilot light fails on my water heater. It works like this:

- The pilot light flame heats the end of the thermocouple.
- There are two wires of different metallurgy, twisted together to form a "junction."
- When the junction is heated, some of the energy of the flame generates a direct electrical current flow of a few volts.
- This voltage is sufficient to open a solenoid valve, permitting gas flow to the pilot light burner.

If the thermocouple malfunctions, you can keep the solenoid valve open with a nine-volt battery. But perhaps this is not one of my better ideas.

One would think that, except for the thermocouple burning out, temperature indication is reliable and may be used with confidence by the Process Control Engineer. After all, the thermocouple wires are protected by the thermowell. This is a thick pipe made of high-grade stainless steel, sealed at the process end. Unfortunately, such temperature indication has a whole range of problems.

Deposits on the surface of the thermowell will insulate the junction of the thermocouple wires. The external portion of the thermowell assembly radiates some heat. The heat loss from the thermowell is normally of no consequence. But if a portion of the thermowell inside the process vessel is fouled, the entire TI assembly will cool. I have observed temperature readings inside vessels operating at 800 °F suppressed by 20–30 °F because of coke formation around the thermowell. To verify this problem, temporarily wrap insulation around the external portion of the thermowell assembly. If the TI reading increases by 5–10 °F, the thermowell is fouled and reliable temperature measurements cannot be determined.

I was working for Exxon on a vacuum tower problem recently. The tower feed temperature was 760 °F. Eight feet above the feed nozzle, in the flash zone, the temperature of the rising vapors was only 680 °F. What happened to the 80 °F? The answer was "Nothing." Above the flash zone thermowell there was the gas oil product draw-off pan. The pan has a drain hole, so that cool liquid accidentally fell onto the thermowell. I checked the vessel's external skin temperature around the entire flash zone. It was all quite uniform and consistent with the 760 °F feed temperature. Any single temperature indication in a large diameter vessel may not mean too much. The Process Control Engineer should specify several TI points at the same elevation. This was the practice for the ten-foot-diameter hydrocracker reactors designed for the Amoco facilities in Texas City.

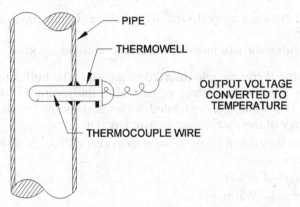

Figure 2-2   *The thermocouple wire should be fully inserted in the thermowell to avoid low temperature indication*

Even in process lines as small as 14 inches, there can be a similar problem. The Texas City hydrocracker reactor feed furnace was overheating, yet the heater outlet temperature was relatively cool. On occasion, a thermowell extending a short distance into the pipe could cause an erroneously low temperature to be observed. However, in this case the thermowell length was fine, but the thermocouple wire was not fully inserted into the thermowell. Figure 2-2 shows a proper installation with the thermocouple wire fully inserted in the thermowell. Pulling out the thermocouple wire to check the depth of the thermowell and the length of the thermocouple wire is safe and easy. But **never ever** unscrew the thermowell while the unit is operating.

FURNACE TEMPERATURES

It is important not to overheat the combustion zone of a heater. The temperature is monitored with thermowells inserted into critical points of the furnace's firebox. We refer to these temperature points as:

- The bridge wall temperature
- The radiant section temperature
- The firebox temperature

Unfortunately, measuring the interior temperature of a firebox with an ordinary thermowell-thermocouple arrangement, as shown in Figure 2-2, is not possible. I was told, in 1965 by my boss, Bill Duval, that the problem was re-radiation of radiant heat from the thermowell. Bill said the problem could be avoided by using a "velocity thermocouple," which draws hot combustion gas past the thermocouple. I've never seen a velocity thermocouple used in any plant, so I will not comment on Mr. Duval's recommendation.

But the problem is very real. One of my hobbies is making ceramics, which I fire in a kiln.

There are three ways to monitor the temperature in my kiln:

1. A portable thermocouple inserted in the kiln. The hotter the temperature, the greater the voltage generated at the junction of the thermocouple. The voltage is correlated with temperature for the particular metallurgy of the thermocouple wire junction.
2. The refractory color. I have a color chart that correlates with the refractory temperature:
   - Dark red—Coolest
   - Cherry red—Warmer
   - Dark orange—Medium hot
   - Light orange—Hotter
   - Yellow—Hottest
3. Clay cones that sag at various temperatures, which I can observe through a sight port. I have 19 sets of these indication cones.

Up to 1000 °F, there is no problem. All three methods track exactly. At 1200 °F the portable temperature indicator lags the other two methods by 50 °F. By 1500 °F the indicator is 100 °F low. By 1800 °F the indicator is 200 °F low. For me, the clay cones are the definitive measurement.

During the 1980 strike in the Amoco Refinery in Texas City, I worked as the outside operator on the sulfur plant. Restarting the sulfur plant furnace, I monitored the furnace temperature with a thermocouple inserted through a ceramic thermowell. Perhaps the thermowell was too short? Maybe the problem was re-radiation, like Bill Duval warned me about. Perhaps the thermowell was encrusted with deposits. I don't really know. But what I do know is that the heater's refractory was glowing a brilliant, blinding yellow color. What I know for sure is that I melted the refractory-lined ends of the boiler tube sheet.

Mr. Durland, the plant manager, joined me for dinner that evening. His only comment about my sorry story was, "optical pyrometer."

Is this fair? Should the responsibilities of the Process Control Engineer include knowing that radiant heat temperatures (1200 °F plus) require the use of an optical pyrometer, not just a thermocouple? Apparently that was Mr. Durland's attitude. Actually, the sulfur plant had an online optical pyrometer. But its little sight-port glass was dirty. It also was reading too low. But, that was not my fault either.

MEASURING PRESSURES

Ladies and gentlemen, what then is the real function of the Process Control Engineer? To worry! My mother would have made a wonderful control engineer. She worried about everything and everyone. Even measuring pressures is worth worrying about. Pressures are measured by creating a strain or deformation on a flexible disk. The deformation of the disk generates an electrical output in proportion to the pressure. There are two possible problems that commonly result in an incorrect pressure signal transmission. These are plugged pressure taps and leaking connections.

In 1974 I almost blew up my alkylation unit depropanizer tower in Texas City. The normal operating set point was 300 psig. The set point pressure was controlled by the reboiler heat input (see Chapter 5, "Distillation Tower Pressure Control"). More heat generated more pressure.

To combat corrosion, a sticky black chemical additive was injected into the reflux flow. This additive partially plugged the pressure connection on the top of the tower. Partly plugged pressure connections will respond slowly to pressure changes but eventually read correctly. Unless they are also leaking. The leak will cause the pressure to read low. The low pressure reading caused the reboiler heat input to increase. The increased reboiler duty and higher temperature drove the sticky chemical additive overhead in greater quantities. This accelerated the rate of plugging, not only of the pressure connection but also of the overhead pressure release safety valve. How did I come to know these things?

One week after initiating the corrosion control chemical injection program I noticed that a local pressure gauge (PG) on the reflux drum (Fig. 2-3) was reading 450 psig. The pressure safety valve was supposed to open at the tower

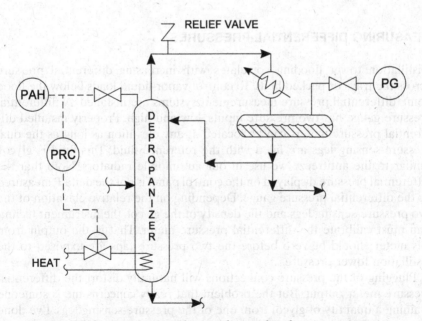

Figure 2-3 *Never connect an alarm and control to the same sensing point*

design pressure of 350 psig. Not only was the pressure tap plugged, so was the relief valve. The Process Control Engineer who specified the instrumentation on the P&ID (Process & Instrumentation Design) indicated that a PAH (Pressure Alarm High) instrument was required.

In case the normal PRC (Pressure Recorder Control) loop malfunctions, this would alert the panel board operator to switch from automatic to manual pressure control of the reboiler. However, if the pressure alarm is located at the same sensing point as the pressure control, as shown in Figure 2-3, then the high pressure alarm will not be activated. Perhaps the worst part of this story is that the pressure on the reflux drum was dropping from an even higher pressure when I noticed the problem. Mechanical failure of a vessel at 50% above its relief valve setting is certainly possible.

In summary, always avoid connecting a control point and an alarm to the same sensing point. This is especially true for a pressure-sensing point when the pressure is controlled by the column heat balance.

Once while working in a refinery in Lithuania I observed that a reflux drum pressure had dropped to a subatmospheric pressure of 10 psia or minus 5 psig. This is dangerous. Too low a pressure may cause a vessel to collapse. The reflux drum had a properly functioning pressure indicator, but the minimum pressure that could be measured was only atmospheric pressure. Hence, the operators were unaware that a dangerous vacuum had developed in the reflux drum. The Process Control Engineer had failed to appreciate that the frigid winters in Lithuania could result in extremely low condensing temperatures and pressures.

MEASURING DIFFERENTIAL PRESSURES

Distillation tower flooding correlates with increasing differential pressure across the trays or packed beds. To control vapor-liquid loads below the flood point, differential pressure measurement systems are installed. A differential pressure gauge has two pressure inputs; low and high. Properly installed differential pressure meters can be located at any elevation as long as the dual pressure-sensing legs are filled with the reference fluid. This fluid is glycol, similar to the antifreeze we use in our automobile radiators. Note that the differential pressure displayed on the control panel is not the delta P measured by the differential pressure gauge. Depending on the relative elevation of the two pressure-sensing legs and the density of the glycol, the instrument technician must calibrate the differential pressure meter. That is, the output from this meter should be zero before the two pressure taps are exposed to the distillation tower pressure.

Plugging of the pressure connections will naturally distort the differential pressure meter output. But the problem that really concerns me is someone draining a quantity of glycol from one of the pressure-sensing legs. I've done this myself. Thinking that a pressure connection is plugged, I'll check the connection by opening a drain valve. Often, differential pressure meter sensing lines are filled not with glycol but with process fluid. In that case, opening a drain will not affect the meter's output as it will refill itself with process fluid. But if I stupidly drain glycol, I've changed the calibration of the differential pressure meter. I should report this to the instrument technician, who would then recharge the pressure-sensing leg with additional glycol. But then I would have to admit my stupid error.

If the Process Control Engineer observes a sudden unexplainable change in distillation tower delta P, the cause may not be a process problem but a loss of fluid from one of the pressure-sensing glycol-filled legs.

HOW ARE FLOWS MEASURED?

"Mr. Lieberman," one of my students asked, 'When are we going to discuss online computer control? How about real-time parameter optimization? You have yet to mention multivariable advanced computer optimization, either. When will we review the application of partial differential equations to Process Control Engineering?"

"Quite outright," I responded, "Let's learn to use math to calculate process flows. First, note that in Figure 2-4 we are measuring delta P, not flows. The

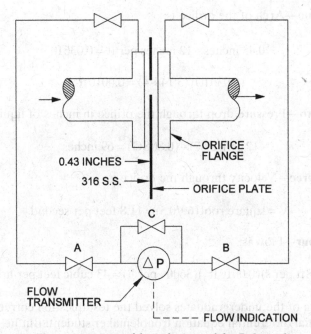

Figure 2-4 *Flows are measured by inducing a delta P through an orifice*

flow transmitter is really a differential pressure measuring device as I discussed before. To convert from delta P to flow we:

- Convert the measured delta P in psi to (delta H), which is inches of liquid pressure drop:

- $\text{Delta H} = \dfrac{(\text{delta P in psi}) \times 27.7}{(\text{fluid specific gravity})}$

Next, calculate V, the velocity through the orifice plate in feet per second:

$$V = \text{square root}(\text{delta H/K})$$

"K" above is the orifice plate coefficient. A typical orifice coefficient is 0.5, but refer to your Crane Hydraulic Data Book for the precise value. Having calculated the velocity through the orifice plate, multiply V by the area of the orifice plate hole. The diameter of the orifice plate hole is stamped on the handle protruding from the orifice plate flanges. As shown in Figure 2-4, this is 0.43 inches.

I tested my class of undergraduates with the following problem:

The above orifice plate has a measured delta P of 2 psi. The fluid has an 0.80 SG.

What is the flow in cubic feet per hour? The solution is:

- **Step One**—Area of the orifice is:

$$0.43 \text{ inches} \div 12 \text{ inches per ft} = 0.036 \text{ ft}$$

$$(0.036 \text{ ft})^2 (3.14 \div 4) = 0.0010 \text{ ft}^2$$

- **Step Two**—Pressure drop through the orifice in inches of liquid is:

$$(2 \text{ psi})(27.7) \div (0.80 \text{ SG}) = 69 \text{ inches}$$

- **Step Three**—Velocity through the orifice is:

$$V = \text{square root}(69/0.5) = 11.8 \text{ feet per second}$$

- **Step Four**—Flow is:

$$(11.8 \text{ ft per s})(0.0010 \text{ ft}^2)(3600 \text{ s per h}) = 43 \text{ cubic feet per hour}$$

Every one of the undergraduates solved the test question correctly, except for that partial differential equation troublemaker student. But he did have a few good questions.

"Mr. Lieberman, suppose the orifice plate is installed backwards. That can easily happen."

"Yes, Mr. Troublemaker. You can tell if the plate is installed backwards because the material designation (316 stainless steel) and orifice size will be facing downstream. The orifice has a beveled edge to reduce fluid turbulence. If the beveled edge is reversed, the orifice coefficient K increases by 20%. And the indicated flow increases by 10% because of increased turbulence."

"Mr. Lieberman; Figure 2-4 shows that the orifice taps are on top of the orifice flanges. Would it not be simpler to connect the flow transmitter to the bottom of the orifice flanges?"

"Yes, Mr. Advanced Computer Control Troublemaker," I answered. "That would be easier, but then fouling deposits would more readily plug the orifice taps, which would result in an erroneous flow indication. Unfortunately, most process installations connect to the bottom set of taps on the orifice flanges, which leads to unending problems in Process Control Technology."

"Mr. Lieberman, a final question. Based on your review of differential pressure transmitters, it seems that the flow transmitter could also be off-zero. How do we correct for this error?"

"Well, Mr. Advanced Multi-Variable Control Troublemaker, let me explain. First, determine the extent of the off-zero error. Referring to Figure 2-4, close valves A and B and open valve C. Let's assume the flow reads 30 ft^3 per hour. The flow should have gone to zero when I equalized the pressure by opening valve C but as you suggested, Mr. Troublemaker, the flow transmitter had not been calibrated correctly.

"Next, open valves A and B, and close valve C. The indicated flow is now 70 ft^3 per hour. What then is the real flow corrected for the meter's miscalibration?" I asked.

"Easy Mr. Lieberman. It's 70 minus 30, or 40 ft^3 per hour," responded the student troublemaker.

"Not quite right," I answered. "Flow is proportional to delta P squared. To correct for the flow meter being off-zero, and remembering that the flow meter is actually a differential pressure transmitter, we proceed as follows:

• **Step One**—Square the indicated flow:

$$(70)^2 = 4900$$

• **Step Two**—Square the amount that the meter is off-zero:

$$(30)^2 = 900$$

• **Step Three**—Take the square root of the difference between Step One and Step Two:

$$4900 - 900 = 4000$$

$$(4000)^{1/2} = 63 \, \text{ft}^3/\text{hour}$$

"There's a big difference between 63 and $40 \, \text{ft}^3$/second. Operators have been killed because Process Control Engineers did not perform this calculation correctly. I'm speaking about 15 contractors in the BP Texas City Refinery, who burned to death in a gasoline inferno.

Look class, before you worry and wonder about the advanced and complex aspects of process control, you must grasp the basics. It's a waste of time trying to optimize process parameters if we cannot measure these parameters with confidence. Garbage in; garbage out."

After class, the Troublemaker complained to Dr. Dillard Smythe, Dean of the Chemical Engineering School at Louisiana State University in Baton Rouge. Dean Smythe said my trial class had been received with hostility by the undergraduate engineers. Half the students felt I was rude; half the students felt my course was not taught at a university level. And the other half didn't like the way I dressed.

I realize this doesn't add up. Regardless, this was the end of my academic career on the faculty of Louisiana State University. The Troublemaker, however, did very well for himself. He made ten million dollars selling subprime mortgages in New Orleans.

3

Dependent and Independent Variables

Humanity is divided rigidly into two classes. Not men and women. Not rich or poor, black, white, Christian, Moslem, young or old. These sorts of classifications are insignificant. The main division that divides mankind is the belief in the Phase Rule. Understanding the Phase Rule conveys the ability to discriminate between independent and dependent variables. Knowledge of the Phase Rule is instinctive in portions of the human race. I've always known the relationship between dependent and independent parameters. However, some engineers and process technicians never understand this relationship.

I'll be using the following terms, which have similar meanings:

- Degrees of freedom
- Independent variables
- Phase rule criteria

There is very little possibility of being a successful Process Control Engineer or panel board operator without a grasp of the relationship between dependent and independent variables.

An example of the Phase Rule is boiling pure water. I live at sea level, where atmospheric pressure is 14.7 psia. Water boils at 212 °F in my kitchen. In specifying pure water I have specified one independent variable, the composition. In specifying atmospheric pressure, I have specified a second independent variable, the pressure.

Troubleshooting Process Plant Control, by Norman P. Lieberman
Copyright © 2009 John Wiley & Sons, Inc.

According to the Phase Rule, the temperature at which the water boils, 212 °F, is now fixed. The boiling point temperature of the water is called a dependent variable. It has been predetermined by the two independent variables—pressure and composition.

I can cause the water to boil at a lower temperature by adding vodka. This changes the composition of the water, which was one of the independent variables. Or I could move to Machu Picchu in Peru at 12,000 feet above sea level to reduce the pressure, which was also one of our independent variables.

CHANGING THE DEGREES OF FREEDOM

Let's say I'll stay home in New Orleans, at sea level. This fixes pressure. I could lower the boiling point of my liquid to 200 °F by adding 10% ethanol plus 1% methanol. Or I could lower the boiling point to 200 °F by adding 5% ethanol plus 3% methanol. This means I have created another independent variable to manipulate (for a total of three parameters) by creating a third degree of freedom. The new degree of freedom is the ratio of the ethanol to the methanol. To summarize, my independent variables are:

- Pressure
- Concentration of water
- Ratio of alcohols

My dependent variables are the liquid's boiling point temperature and the composition of the vapor generated from the boiling liquid.

VARIABLES IN DISTILLATION

In a simple distillation column (Fig. 3-1) with two pure components in the feed, we can select three independent variables from the following list:

- Tower pressure
- Reflux rate
- Reboiler duty
- Condenser duty
- Any composition on one particular tray
- Composition of the overhead product
- Composition of the bottoms product
- Ratio of overhead to bottoms product
- Tower top temperature

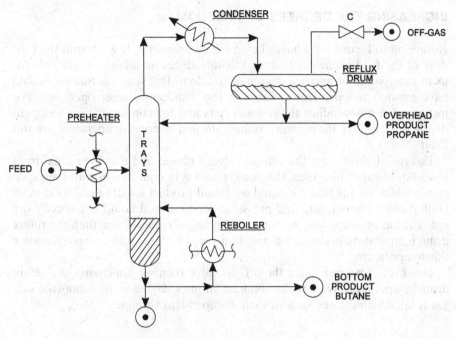

Figure 3-1 *Simple distillation tower*

• Tower bottoms temperature
• Any individual tray temperature

The reason I can select three independent variables in this distillation example rather than just two (as when I boiled water in my kitchen) is that we have introduced another degree of freedom. That additional degree of freedom is tower top reflux.

A typical selection of the three independent variables is:

• Tower top pressure
• Reflux rate
• Tower bottoms temperature

I am not suggesting that this is the optimum selection. It depends on individual circumstances, which I'll discuss later. However, these three independent control targets having been selected, every other parameter listed above, in accordance with the Phase Rule, is now a dependent variable. We are prohibited by the laws of nature from altering any of the many dependent tower operating parameters.

INCREASING THE DEGREES OF FREEDOM

Returning to Figure 3-1, I have shown a feed preheater. If we permit the heat duty of the feed preheater to vary, this introduces an additional variable for us to manipulate or another degree of freedom. This suggests that we should have another independent variable in the distillation tower operation. For instance, I can now adjust the reboiler duty and the condenser duty independently, even though the pressure, reflux rate, and bottoms temperature are still fixed.

Figure 3-1 shows that the off-gas valve is closed and that the reflux drum has a liquid-vapor interface. The liquid phase is in equilibrium with the vapor phase, which means that the liquid overhead product is saturated liquid at its boiling point temperature and pressure. For saturated liquid, if I specify the reflux drum pressure and the reflux drum liquid composition, then the reflux drum temperature becomes a dependent variable. You and I cannot then alter this temperature.

However, if we now open the off-gas valve then we can change the reflux drum temperature. We have introduced another degree of freedom (the off-gas rate), which creates an additional independent variable.

COMPLEX DISTILLATION TOWERS

In a refinery we have crude oil, coking, and cracking distillation towers that are far more complex than Figure 3-1. The sort of towers we might have include:

- Multiple pumparounds for heat removal
- Dual overhead condensing systems
- Side product draw-offs
- Stripping steam injection
- Intermediate naphtha feeds
- Cold pump-downs, also for heat extraction
- Vapor returns from strippers

The ability to grasp which parameters to select independently, and which parameters are consequently dependent variables, comes with experience and training. I rather hate to admit this, but I learned about selecting independent variables by generating computer models for complex coker fractionators. This was in 1965, when we were still using punch cards and Fortran. While I find it painful to admit, the best way for the Process Control Engineer to study the relationship between operating variables is to work with a computer model.

The problem with using computer models to study interaction of process parameters is that it's all steady state. The dynamic feature of the process response is lost. The dynamic nature of the interaction of independent and dependent variables can only be learned by working with the panel board operator. The really experienced and the best panel operators comprehend that process moves cannot be made independently. Every action has consequences.

AMOCO-TEXAS CITY—1974

The idea that actions have consequences is not universally understood. At least my operators in Texas City did not seem to be aware of this principle. On my alkylation unit depropanizer (Fig. 3-1) I had a 2% butane spec for the overhead propane LPG product. The operators understood that more reflux was needed to wash back the heavier butane from the up-flowing vapor. What they did not understand was that the reflux rate is a dependent variable. It depends on the reboiler duty. At steady state it is not possible to increase the reflux rate with a constant reboiler duty. With a constant reboiler duty, the liquid level in the reflux drum will drop and the overhead propane production will rapidly decline. After 10 minutes the reflux pump will lose suction pressure and the pump will cavitate.

I explained repeatedly to my operators that the reflux comes from the reboiler. It seemed so obvious. The flow of vapor up the tower condenses and becomes the reflux. I was puzzled that half the panel men never grasped the relationship between reboiler duty and reflux rate. Thirty-four years later I'm no longer puzzled. Panel operators frequently fail to appreciate the consequences of their actions. That's why the most important job of the Process Control Engineer is one of communications—communicating control concepts to the panel board operator. The most important and difficult concept to explain is the relationship between independent and dependent variables. This includes the concept that the number of independent variables is a function of the degrees of freedom. Use of a computer model is a reasonable and productive method to aid in such an explanation.

VARIABLES IN GAS COMPRESSION

My understanding of the interrelationship of process variables has often been poor until someone else explained the relevant concepts to me. For instance, my first project involving a compressor was for a refrigeration unit in a refinery in El Dorado, Arkansas:

- Compressor type—reciprocating
- Refrigerant—mixture of propane and butane
- Molecular weight—variable

- Suction pressure—50 psia
- Discharge pressure—200 psia
- Refrigerant flow—10,000 lbs/h
- Driver—3600 rpm electric motor

The refrigeration unit design was a success. My supervisor was so pleased that he assigned me to duplicate the design for a new facility in Sugar Creek, Missouri. The only alteration was to be that the compressor would be centrifugal rather than reciprocating. The controls and instrumentation I specified were the same for both compressors. But the result was not another successful design but a failure. The problem was the variable molecular weight of the refrigerant. Whenever the refrigerant molecular weight went up, the suction pressure would fall in a most alarming fashion (see Fig. 3-2). Sometimes the suction pressure would fall below atmospheric pressure, which is dangerous, in the sense that air could be drawn into the process through leaks in the upstream equipment.

I complained to John Houseman, the Senior Rotating Equipment Engineer, that the centrifugal compressor was malfunctioning.

"John, the suction pressure is falling without changing any control valve position. Also, the amperage load on the electric motor driver becomes excessive, which trips off the circuit breaker because of amperage overload. All of this happens without the panel board operator changing any process variable. John, I believe the centrifugal compressor is malfunctioning and needs replacement."

"Lieberman, it is your brain that is malfunctioning and needs replacement," answered Mr. Houseman. "What happens to the molecular weight of the circulating refrigerant when the electric current flow increases and the suction pressure drops?"

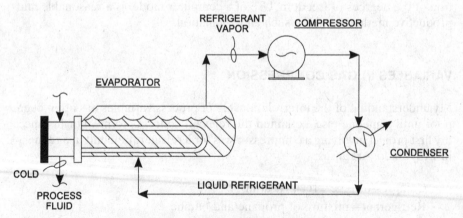

Figure 3-2   *Simple refrigeration circuit*

"I think the molecular weight increases, Mr. Houseman. But I'm interested only in the compressor performance and not the composition of the refrigerant."

"No, Lieberman, you don't understand. Centrifugal compressors running at a constant speed, at a certain flow rate, must produce a fixed amount of polytropic head. To convert from polytropic head to differential pressure you have to multiply by the density of the vaporized refrigerant [see Chapter 15, "Centrifugal Compressors Surge and Over-amping"]. This vapor density is a function of the molecular weight of the gas compressed. Since the compressor discharge pressure is a function of the refrigerant condenser temperature, which is constant, the compressor suction pressure is drawn down. It's all due to the dynamic nature of the centrifugal compressor. It will produce more differential pressure as the molecular weight increases. And this bigger delta P increases the compression ratio. Which all means the amperage load on the motor driver must go up."

"You see, Lieberman, actions have consequences," continued John Houseman. "When you switched from a reciprocating compressor to a centrifugal compressor you failed to realize that you also had to change the Process Control Logic. The variable molecular weight introduces an additional degree of freedom into the centrifugal compressor's operating characteristics. You can't just introduce a degree of freedom without adding an additional control loop into the process. In your case, a new suction throttle valve would be best."

"I'm sorry, Mr. Houseman, but I still don't understand. Why didn't I need the suction throttle valve when I had the reciprocating compressor in El Dorado, Arkansas? It also had to contend with a variable refrigerant molecular weight," I asked.

"Because, Lieberman, the reciprocating compressor doesn't produce a variable differential pressure with a variable molecular weight gas. You introduced the new degree of freedom not by changing the process, but by changing from a positive displacement compressor to a dynamic compressor. Look, Mr. Houseman concluded, "you can't design process controls and specify instrumentation unless you understand how the equipment works. In this case you cannot arbitrarily specify the differential pressure developed by the centrifugal compressor. This delta P is a dependent variable. It depends upon the density or the molecular weight of the circulating refrigerant. The composition of the refrigerant shown in Figure 3-2 is the independent variable, and the compressor suction pressure is the dependent variable."

More or less, I had no idea what Mr. John Houseman was talking about. Just like my operators in Texas City, I didn't understand how to control the equipment because I failed to grasp how the equipment worked. As time progressed, I did learn about centrifugal compressors. In Chapter 15 I review this complex and difficult subject in detail and explain how an increase in gas density at the compressor suction also increases the compressor differential pressure.

We dealt with the problem at the new plant in Sugar Creek, Missouri by reversing the relationship between the dependent and independent variables. We added connections to allow the panel board operator to manipulate the makeup of propane or butane in the circulating refrigerant. If the delta P and the motor amperage current was excessive, the operator would reduce the molecular weight of the refrigerant by spiking it with lighter propane from the LPG tank.

John Houseman and the polypropylene plants in El Dorado and Sugar Creek are gone. The viscous polypropylene was used as a chemical additive in gasoline in the 1960s by Texaco. It probably did more harm than good. The only thing left is my recollection of the friendliness and kindness of the operators at the Amoco Refinery in El Dorado, Arkansas and Mr. John Houseman's patience with a very confused young engineer.

4

Binary Distillation of Pure Components

Water boils at 212 °F at 14.7 psia, provided that we are boiling pure water. But, in a practical sense, the boiling point of water does not change if there are a few thousand ppm (parts per million) of NH_3 or H_2S in the water. Water with 20 ppm of NH_3 boils at almost the same temperature as water with 200 ppm of NH_3. The tower shown in Figure 4-1 is a sour water stripper. It strips NH_3 in the feed water from 20,000 ppm (2%) down to 20 ppm. Steam flowing through tray 1 to tray 10 does the work of stripping. More steam strips harder and reduces the NH_3 content on the stripper bottoms.

The question for the Process Control Engineer is how to control three variables:

- The tower pressure
- The reflux rate
- The reboiler steam rate

The tower pressure is controlled by the NH_3-rich off-gas pressure. This off-gas flows to an incinerator. It's back pressure from the incinerator that controls the stripper pressure.

The tower is on total reflux. There is no net liquid product produced from the reflux drum. The reflux rate varies to hold the level in the reflux drum. That is, the reflux rate depends on the amount of steam being condensed to water in the overhead condenser.

Troubleshooting Process Plant Control, by Norman P. Lieberman
Copyright © 2009 John Wiley & Sons, Inc.

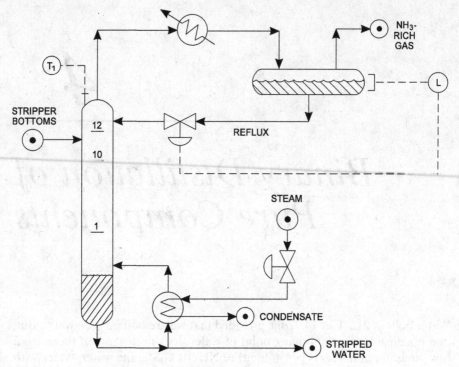

Figure 4-1 *Controlling steam flow to a stripper reboiler*

The steam flow to the condenser is largely a function of the stripping steam generated by the reboiler. To summarize, the reflux rate is controlled by the reboiler duty. But how do we control the reboiler duty?

For an ordinary distillation tower this is not a problem. If the two components are propane and butane, we might have a propane spec of 3% in the butane. If the tower pressure was 300 psig, the tower bottoms temperature would be 220 °F. If the tower's bottom composition was off-spec at 6% propane, the tower bottoms temperature would be 210 °F. The panel board operator would increase the TRC (temperature recorder control) set point from 210 °F to 220 °F. The reboiler duty would increase and, the propane content of the butane bottom product would fall from 6% to 3%.

For the sour water stripper shown in Figure 4-1, the problem is more difficult. Let us assume that the tower operating pressure is 15 psig. The boiling point of water with 200 ppm of NH_3 at 15 psig is about 230 °F. Perhaps the boiling point of water with 20 ppm of NH_3 is 230.1 °F. A temperature controller is not capable of discriminating between 230 °F and 230.1 °F. Even if this were possible, small pressure changes would obscure the effect of temperature. There is simply not enough correlation between composition and temperature to use the tower bottoms temperature to control the reboiler steam flow.

Stripping aqueous streams is a common process operation, not only to strip out NH_3 from water, but also to strip:

- H_2S from amine solutions
- Phenols from refinery waste water
- Benzene from seawater used as ballast in tankers
- Alcohols from chemical plant waste water
- Naphtha from extraction water used in tar sands

Most of my experience is in stripping H_2S from circulating amine-water solutions. A few hours as a panel operator on such a system would prove that controlling reboiler steam flow on closed-loop automatic temperature control is futile. The steam flow rate would prove to be entirely unstable.

One common option to restore reboiler steam flow stability is to use T_1, the tower top temperature, to control the reboiler steam flow. This option in effect controls the ratio of NH_3 to H_2O distilled overhead. The temperature of the vapor leaving the top tray is the same as the boiling point temperature of the liquid on the top tray. This liquid is rich in NH_3. As the ratio of NH_3 to H_2O changes at the top tray, there is, compared to the bottom tray, a big change in tower top temperature. This gives the tower top TRC a substantial temperature gradient to use in controlling the reboiler steam flow. As a positive consequence, the reboiler steam flow is stable, and so is the tower top reflux rate.

Looks can be deceiving. Most console operators observing the stable operation of the sour water stripper, controlled by the top TRC manipulating the heat flow to the reboiler, would be well satisfied. Process Control Engineers working with a computer simulation model would agree that the top TRC controlling steam to the reboiler ought to work fine. Unfortunately, in practice the result is bad. Whether we are stripping H_2S from amine or NH_3 from sour water, the residual NH_3 or H_2S in the bottoms will be erratically high. Of course, unless there is an on-stream, continuous analyzer, this erratic parameter remains unknown to the panel operator.

The problem is that a small increase in tower operating pressure will radically alter the ratio of NH_3 to H_2O on tray 12, compared to tray 1. Variations in the feed temperature would also affect tray 12 far more than tray 1 temperature (Figure 4-1).

PROPER CONTROL OF WATER STRIPPERS

Based on long experience, the ancient engineers Larry Kunkel and Howard Krekel at the long forgotten Pan American Natural Gas Corporation found that reboiler steam flow should best be controlled on "Feed Forward Ratio Control." By experiment they found that properly stripped amine required 0.9

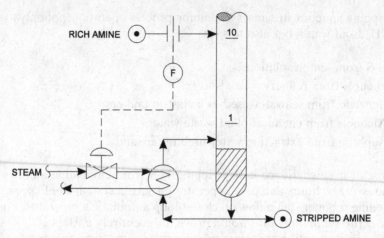

Figure 4-2 *Rich amine flow controls steam to the reboiler to hold a fixed ratio of steam to amine*

to 1.3 pounds of reboiler steam per gallon (U.S.) of stripper feed. The control scheme they used, shown in Figure 4-2, worked rather well in gas field operations, where conditions are quite consistent and stable.

In my refinery applications, the results were less satisfactory. The problem was that the rich amine feed temperature was variable. In a refinery the rich amine feed to the stripper originated from a dozen sources. As the feed temperature declined by 12 °F, an increase in reboiler duty of 0.1 pounds of steam per gallon of feed was needed. This extra steam did not improve stripping.

It was just the extra latent heat of steam needed to offset the lost sensible heat due to the colder feed. I've provided the calculations:

$$(12°F) \cdot (1.0\,BTU/°F/lb) \cdot (8.34\,lb/gal) \div 980\,BTU/lb\ steam = 0.1\,lb$$
$$(i.e.,\ pounds\ of\ extra\ steam\ per\ gallon\ of\ feed)$$

Currently, with computer technology, an online correction factor to the steam ratio controller for feed temperature could easily be implemented. But I've never done this in practice.

What I have done is to control the reboiler steam flow based on maintaining a reflux ratio target. Let me explain:

- Some of the reboiler duty heats the stripper feed (see Fig. 4-3) from 190 °F to 250 °F.
- Some of the heat duty of the reboiler breaks the chemical bond between the absorbed gas (NH₃ or H₂S) and the solvent (amine or water).
- The rest of the heat duty of the reboiler generates stripping steam. This steam flows through the stripping trays to do the work of stripping.

• The internal stripping steam flowing through the trays goes overhead. It condenses to generate tower top reflux flow.

In summary, what I want to control in the stripper is the ratio of steam flow through the stripping trays divided by the feed flow. To accomplish this objective I set the reflux rate to hold a ratio of feed to reflux. This ratio then controls the flow of steam to the reboiler.

Certainly I have just proved that one picture (Fig. 4-3) is worth a thousand words. Study my sketch and reread the above section, because even my mentor, Larry Kunkel, would need a second reading.

Champlin Oil is another one of my clients that has expired. They operated a tower at their Corpus Christi Refinery designed to produce propylene. The

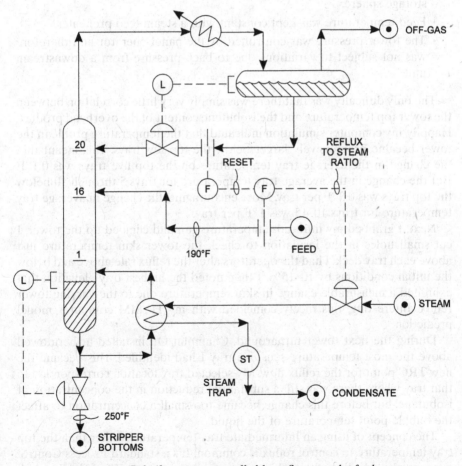

Figure 4-3 Reboiler steam controlled by reflux, reset by feed rate

propylene was reacted with benzene to make pseudo-cumene, which ultimately was turned into plastic car bodies.

The propylene stream had to be carefully distilled to meet a residual isobutene specification of not more than 0.1%. At a tower pressure of 320 psig the boiling point of 99.9% pure propylene and 0.1% isobutene is 121 °F. At a tower pressure of 320 psig the boiling point of 99.7% pure propylene and 0.3% isobutene is 121.2 °F. Temperature controllers cannot discriminate between 121 °F and 121.2 °F. Even if they could, minor changes in the tower pressure would obscure the 0.2 °F difference.

It rather seems like I'm repeating the story just related about reboiler steam control to my sour water stripper. The only difference is that here we are concerned with controlling the reflux rate rather than the reboiler duty. In a way, this is simpler than the water stripper because:

- Feed rate was kept absolutely constant by flow controlling out of a large storage sphere.
- Feed temperature was kept constant with a steam feed preheater.
- The tower pressure was controlled by the panel operator and therefore was not subject to variations due to back pressure from a downstream unit.

The only difficulty was that there was simply very little correlation between the tower top temperature and the isobutene content of the overhead product. Happily, my computer simulation indicated that the temperature profile in the tower became progressively larger below the top few trays. This meant that the change in the average tray temperature on the top five trays was 0.1 °F. But the change in the average tray temperature for trays 5 through 10 below the top trays was 0.3 °F per tray. More encouraging, the change in average tray temperature for trays 10–15 was 1 °F per tray.

Next, I grabbed my infrared temperature gun and climbed up the tower. I cut small holes in the insulation to check the tower skin temperature just above each tray deck. I had the operators alter the reflux rate above and below the initial conditions by 10–15%. I then noted the highest tray elevation that exhibited a measurable change in skin temperature due to the varying tower reflux. Interesting, this nicely coincided with my HYSIM computer model prediction.

During the next tower turnaround, Champlin Oil installed a thermowell above the most temperature sensitive tray I had identified. This became the new TRC point for the reflux flow. My selected tray location corresponded to that tray where there was still a substantial reduction in the concentration of isobutene, but before this change became too small a concentration to affect the bubble point temperature of the liquid.

The concept of using an intermediate tray temperature rather than the top tray temperature to control reflux is common. It's a standard Process Control concept when dealing with high-purity overhead products. When dealing with high-purity bottoms products the same concept is applied to the reboiler duty, except that an intermediate tray below the feed point is used, rather than the tower bottom product temperature or the reboiler outlet temperature.

My only contribution to this common practice was my method of selection of the optimum tray location, that method being the combination of field observations integrated with my tray-to-tray computer model of the fractionation column. But even this idea I learned from Howard Krekel, the Pan American patriarch.

5

Distillation Tower Pressure Control

Distillation towers are at the heart of any process plant or petroleum refinery. Fractionation efficiency in all distillation towers is quite impossible with an erratic operating pressure. Thus, the critical importance of precise tower pressure control.

I once had a job for Koch-Glitsch, the major worldwide tray manufacturer to troubleshoot a de-isohexanizer tower in Canada. The Canadian refinery claimed the fractionation trays supplied by Koch-Glitsch were developing poor tray efficiency. They were correct. However, the low tray efficiency was not a consequence of inadequate tray design. The poor tray efficiency was a function of the tower's pressure instability.

The liquid on each tray is saturated liquid at its boiling point or bubble point. When the tower's pressure suddenly drops by one or two percent, the liquid on each tray begins to boil and froth, quite similar to an agitated bottle of warm beer that is suddenly opened and depressured. The resulting high froth levels on each tray promote the entrainment of heavier components which contaminate the overhead distillate product.

On the other hand, when the tower pressure suddenly rises by a few percent of operating pressure, the vapor flow through the tray decks rapidly slows for a few seconds. The pressure drop of the vapor flowing through the valve caps or sieve holes also momentarily falls. This can cause the tray to dump, or leak, or weep liquid through the tray deck. Tray deck weeping is the major cause of reduced vapor-liquid contacting and tray efficiency. The weeping trays will contaminate the bottoms product with lighter components.

Troubleshooting Process Plant Control, by Norman P. Lieberman
Copyright © 2009 John Wiley & Sons, Inc.

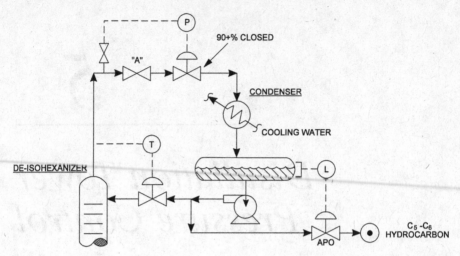

Figure 5-1   *Pressure fluctuations due to control valve operating in its nonlinear range ruin efficiency*

The pressure control scheme for the Canadian de-isohexanizer is shown in Figure 5-1. The back pressure control valve was in a mostly closed position. As I describe in Chapter 10, "Sizing Control Valves," this control valve was operating in a nonlinear portion of its range of control. This means that a small change in the control valve position resulted in a large change in the de-isohexanizer tower pressure. To rectify this situation I slowly closed isolation gate valve "A" upstream of the back pressure control valve until this control valve was open between 50% and 60%. The control valve was now operating in the linear portion of its range of control. The tower pressure variations diminished. Fractionation efficiency was restored to an acceptable level.

The sort of pressure control shown in Figure 5-1 is not good design practice and has a number of disadvantages:

1. For any control valve to control, it must have a pressure drop. This means that the condenser operating pressure will be significantly less than the tower pressure. The overhead product condensing temperature will also be reduced, thus making it more difficult, perhaps impossible, to fully condense the overhead vapor flow.

2. Depending on the cooling water supply temperature to the condenser, the reflux drum pressure may be extremely variable. This can cause the back pressure control valve to operate either too far open or too far closed for proper pressure control.

3. Because there is a valve in the line between the tower and the reflux drum, these two vessels can be isolated from each other. This means that each vessel must be independently protected from excessive pressure by

its own dedicated relief valve. If there were no valves in the overhead vapor line, only a relief valve on the tower top would be required. Relief valves and their piping connections to the flare system are expensive.

4. The liquid in the reflux drum will never be subcooled. This will minimize the net positive suction head (NPSH) available for the reflux pump (See Chapter 16, "Control of Centrifugal Pumps").

COOLING WATER THROTTLING

Placing a control valve on the cooling supply as shown in Figure 5-2 is effective in providing stable distillation tower pressure control. Also, there is no control valve pressure loss between the tower and the condenser. Furthermore, only a single relief valve is needed to protect both the reflux drum and the distillation column. Reducing cooling water flow will reduce the condenser duty. This will increase the temperature in the reflux drum. The vapor pressure of the liquid in the reflux drum will increase. This increases the reflux drum pressure, and the tower pressure simply floats on the reflux drum pressure. Unfortunately, the reduced cooling water flow raises the water outlet temperature above 120–130 °F. Depending on the cooling water quality, calcium carbonate salts will precipitate as hardness deposit inside the tubes. Gradually the heat transfer efficiency of the condenser is severely reduced. For this reason pressure control by throttling on the cooling water is no longer widely accepted in the hydrocarbon industry.

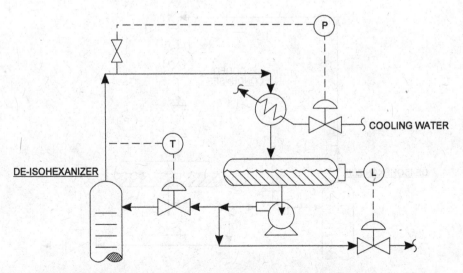

Figure 5-2 *Control of cooling water flow for pressure control is not recommended*

For air-cooled condensers the problem is effective control of the air flow. Louvers placed on top of the air cooler bundle are mechanically unreliable. Variable fan blade pitch or variable-speed motor drivers would help, but over a limited range. For example, what happens when the fan is shut off at a time of low feed rates or cold ambient air conditions?

PRESSURE CONTROL WITH NONCONDENSABLE VAPORS

If vapor must be vented from the reflux drum, there is only one practical method of tower pressure control, that is, a back pressure control valve on the reflux drum off-gas vent line. Sometimes the amount of vapor to be vented is extremely small, or very variable, or it could drop to zero. This greatly increases the complexity of tower pressure control. The correct method of pressure control, when the flow of noncondensables may start and stop, is discussed in a subsequent section of this chapter. As a way of introducing this subject, let us consider flooded condenser control as shown in Figure 5-3.

FLOODED CONDENSER PRESSURE CONTROL

The tower pressure control method depicted in Figure 5-3 represents improper design. From the process control perspective it works fine. When the tower pressure is too low, the control valve draining the condenser closes. The liquid

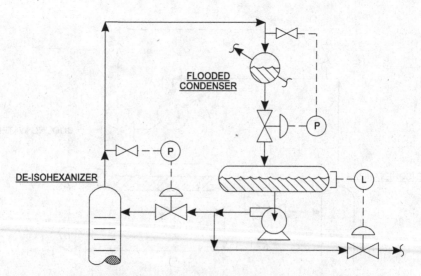

Figure 5-3 *Flooded condenser pressure control incorrectly applied*

level in the condenser rises and reduces the tube surface area exposed to the condensing vapor. The rate of vapor condensation falls, and thus the tower pressure increases to restore the set point pressure. As long as the pressure control valve is in its operable range, stability is good. The problems with this sort of flooded condenser pressure controls are:

- A relief valve would be required on both the reflux drum and tower because the two vessels may be isolated from each other by closing the pressure control valve. This is a requirement of your state boiler code.
- The pressure control valve introduces a pressure drop between the condenser and the reflux drum. Even when the control valve is 100% open, there will still be a delta P across the valve. The reduced reflux drum pressure will promote the evolution of vapor in the reflux drum. The uncondensed vapor would then be vented off to the plant flare or fuel gas system. In effect, condenser capacity is reduced by the control valve located at the condenser outlet line.
- The liquid in the reflux drum is at its bubble point. This results in a minimum net positive suction head available to the reflux pump.

The correct method to use for flooded condenser pressure control is illustrated in Figure 5-4. The liquid level in the condenser rises because of condensate backup so as to increase the tower pressure. The relative elevation of the drum and condenser is irrelevant. The method works well for both shell and tube water condensers, and for air-cooled fin-fan condensers. Not only does this mode of control provide a stable tower pressure, but it also has the following advantages:

- There is no control valve between the tower and the reflux drum. This maximizes both condenser capacity and the reflux drum pressure.
- It's the simplest scheme, as only a single control valve is used to control tower operating pressure.
- The reflux drum liquid is subcooled. This provides more available net positive suction head to the reflux pump than if a level was held in the reflux drum.
- Only a single relief valve is required on the tower and none on the drum, as the two vessels cannot be isolated from each other.

The single drawback of the flooded control scheme shown in Figure 5-4 is that the vent on top of the reflux drum has to be a tight shut-off, leak-proof type of control valve. If it does not seat tightly when closed, liquid rather than vapor will pass. The rate of liquid passing will be roughly 100 times as great as a vapor flow would be through the same leaking valve.

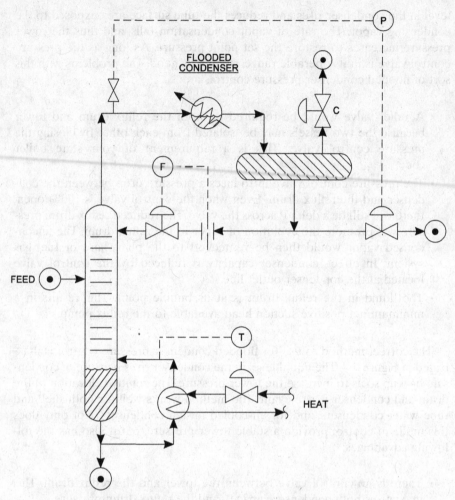

Figure 5-4 *Flooded condenser pressure control is correct design practice*

HOT VAPOR BYPASS PRESSURE CONTROL

Unfortunately, the common alternate to flooded pressure control in wide-spread use is hot vapor bypass control, as shown in Figure 5-5. This inferior method of control works to raise the tower pressure as follows:

- The hot vapor bypass valve, which bypasses the condenser, opens to increase the temperature in the reflux drum.
- The contents of the reflux drum are not cold enough to absorb all of the vapor entering the reflux drum.

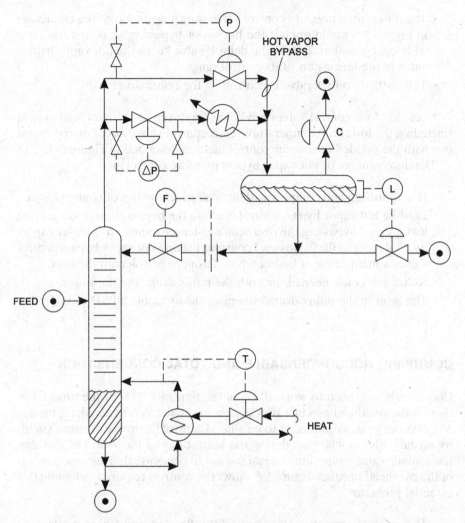

Figure 5-5 *Hot vapor bypass pressure control is wrong design practice*

- The uncondensed vapors accumulating in the reflux drum raise the drum pressure.
- The higher drum pressure pushes up the liquid level in the condenser.
- With fewer tubes exposed to the vapor, the rate of vapor condensation falls.
- The slower rate of vapor condensation raises the tower operating pressure to the required set point.

- The differential pressure control valve shown upstream of the condenser in Figure 5-5 would open if the hot vapor bypass valve is in a too far a wide open position. That is, the delta P valve keeps the hot vapor bypass valve in the linear part of its control range.
- The distillate outlet valve is controlling the reflux drum level.

Note that four control valves are being employed in the overhead system (including the tower top temperature reflux control valve) as compared to just two with the flooded condenser control method described in Figure 5-4.

The disadvantage of hot vapor bypass pressure control are:

- It is confusing and complex with an excessive number of control loops.
- I dislike hot vapor bypass control because the bypass control valve often leaks. Vapor bypassing the overhead condenser reduces condenser capacity. This leads to flaring losses. I consider leaking hot vapor bypass control valves a main cause of hydrocarbon flaring in petroleum refineries.
- Relief valves are needed for both the reflux drum and the tower.
- The level in the reflux drum minimizes the available NPSH to the reflux pump.

COMBINING NONCONDENSABLE AND TOTAL CONDENSATION

Control schemes need to work 100% of the time, not 99% of the time. Even though the overhead product of a distillation tower is fully condensable for 364 days per year, we still need to be able to control the tower pressure stability on July 30th at 4:00 p.m. during the hottest day of the year. On that day, the cooling water temperature is just too hot to condense that last one percent of the overhead product. Figure 5-6 shows the controls required to handle this essential problem:

- The normal operation is identical to flooded condenser control as described above. The overhead product valve (labeled "C") will open to reduce the liquid level in the condenser. This exposes more tubes to the condensing vapor to accelerate condensation and thus reduce the tower pressure back to its set point.
- Instrument air signals flowing through the two connections labeled "B" are not in use as long as the indicated level in the reflux drum is 100%. The instrument air signal flowing through the connection labeled "A" is controlling tower pressure.
- Condenser capacity, because of hot weather, gradually becomes insufficient to condense the entire overhead product. The liquid level in the condenser falls to zero. As valve "C" opens, the level in the reflux drum drops from 100% to 90%.

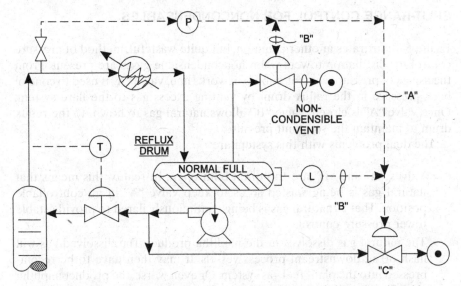

Figure 5-6 *Flooded condenser pressure control with provision for noncondensables*

- The instrument air signal flowing through connection "A" is now disregarded by valve "C." This valve is now controlled by the reflux drum level indicated by the instrument air signal flowing through connection "B."
- The tower pressure is now controlled by the noncondensable vapor vent at the top of the reflux drum by the instrument air signal flowing through connection "B" from the tower top pressure transmitter shown in Figure 5-6.

The tower is now in conventional off-gas vent pressure control, with the reflux drum on ordinary level control. This is accomplished by selecting various instrument air pressures to operate certain control valves.

The control system is set up so that a designated control valve will only be manipulated by a certain range of instrument air pressures. This type of control is called "split-range pressure control" (see Chapter 12).

To summarize, the overhead pressure control system switches automatically between vent control and flooded condenser control, based on the output from the level indicator in the reflux drum. If the level is at 100%, pressure control is by valve "C." If the level in the reflux drum is less than full, pressure control is maintained by the noncondensable vent. All this is done smoothly, easily, and reliably by split range control. It sounds complex, but it's all conventional.

SPLIT-RANGE CONTROL FOR NONCONDENSABLES

Figure 5-7 illustrates another common, but quite wasteful, method of pressure control for distillation towers when noncondensable gases are present. From the aspect of pressure control stability it works fine. Valve "A" is used to control back pressure in the reflux drum by venting excess gas to the flare system. Once valve "A" is closed, valve "B" allows natural gas to flow into the reflux drum to maintain the set point pressure.

The dual problems with this system are:

- Valves "A" and "B" are too often both open. In reality, this means that natural gas is being wasted so as to keep valve "A" in an controllable position. That is, natural gas is being continuously flared to provide stable tower pressure control.
- The natural gas dissolves in the naphtha product. The dissolved gas will flash off in downstream process vessels. It may then have to be recompressed into the plant fuel gas system. Or even worse, the product naphtha may flow into a storage tank. The dissolved natural gas will flash off through the tank's atmospheric vent. The evolved gas will contain significant quantities of light naphtha, which will also be lost to the environment, thereby promoting additional greenhouse gases.

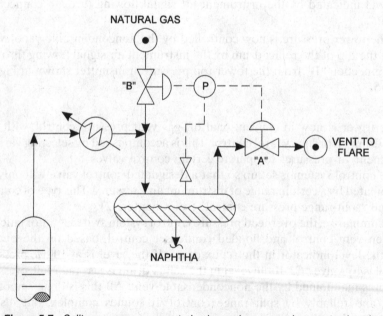

Figure 5-7   *Split-range pressure control using makeup gas when vent valve shuts*

Split-range distillation tower pressure control has been used by major engineering contractors during my entire 44 year career in the refining industry. In 1965, natural gas was essentially free and no one cared about greenhouse gases. In today's environment no reputable Process Control Engineer should use this archaic method of distillation tower pressure control.

REBOILER CONTROLS TOWER PRESSURE

Controlling tower pressure by the reboiler duty, as shown in Figure 5-8, is an alternative and quite acceptable variation to flooded condenser pressure

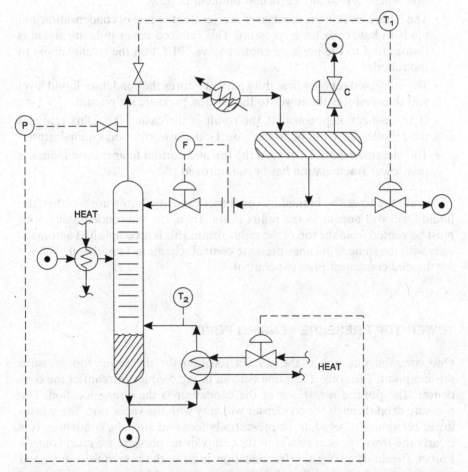

Figure 5-8 *Head input directly controlling tower pressure is correct design practice equivalent to flooded condenser pressure control*

control. Increasing the reboiler duty raises the tower pressure. In 1974 all my towers on my alkylation plant in Texas City were converted to pressure control using the reboiler, and they worked fine. The control scheme functioned as follows:

- Let's assume the panel board operator wants to improve fractionation between the overhead and bottoms. She would then raise the reflux rate, which is on flow control (FRC).
- The tower top temperature would start to fall with the larger reflux flow. The overhead product temperature control valve (TRC) would open to restore the tower top temperature at T_1.
- Opening this temperature control valve reduces the level in the overhead condenser. Note that we do not measure the condenser level, nor do we care where it is at any particular moment in time.
- The falling level in the condenser accelerates the rate of condensation and tends to lower the tower pressure. This reduced tower pressure signal is transmitted to the pressure control valve (PRC) on the steam supply to the reboiler.
- The increased reboiler heat duty partly restores the condenser liquid level and thus restores the tower to the desired pressure set point.
- If T_1 has been kept constant, the result of increasing the reflux rate and the reboiler duty will be to increase T_2, the tower bottoms temperature.
- The increased delta T between the top and bottom temperature indicates that tower fractionation has been improved.

Should the tower be limited by condenser capacity during hot weather, the liquid level will appear in the reflux drum. Then, the noncondensable vapor must be vented from the top of the reflux drum. This is accomplished automatically with the same split-range pressure control scheme as I just detailed above for flooded condenser pressure control.

TOWER TOP PRESSURE SENSING POINT

One common question is the correct location for the tower top pressure sensing point. The correct location (shown in Fig. 5-8) is upstream of the condenser. The point downstream of the condenser is the wrong location. The pressure drop through the condenser will vary with the vapor rate. There is no linear relationship between the pressures before and after the condenser. It is clearly the tower pressure and not the reflux drum pressure we must control. Pressure stability above the fractionation trays is the prerequisite for good tray efficiency. Not more than 5–10% of the towers I have seen have located the pressure sensing point downstream of the condenser.

If you have a tower with a pressure control point on the reflux drum, or at the condenser outlet, I would advise you to relocate the pressure sensing point to the condenser inlet or, even better, at the very top of the column. This should result in both improved tower pressure stability and enhanced tower fractionation efficiency.

6

Pressure Control in Multicomponent Systems

Heat makes pressure. This concept is central to the process control of distillation towers, as discussed in Chapter 5, "Distillation Tower Pressure Control." However, there are no real definite rules in process control engineering. Sometimes increased reboiler duty will reduce pressure in a distillation tower. The console operators at the Texaco Refinery in Port Arthur, Texas proved this by example.

GETTING A SAMPLE IN A BOTTLE

Steve, the plant control engineer, was assigned to design a control scheme for the gasoline stabilizer shown in Figure 6-1. He obtained a sample of the tower feed at the indicated sample point in the figure in a bottle. As the sample he drew was at atmospheric pressure and the hydrocarbon in the feed drum was at 30 psig pressure, the liquid flashed as it entered the bottle. Steve, having failed to notice the loss of light hydrocarbon vapor, submitted the sample to the laboratory for gas chromatographic analysis (GC). The lab result did not reflect the actual amount of methane and ethane in the gasoline stabilizer feed because these lighter components had flashed off. Steve then calculated that the propane and butane in the tower's overhead product would fully condense at 150 psig and 100 °F in the reflux drum as shown in Figure 6-1. Steve's design had no provision for a noncondensable vent from the reflux drum, except to

Troubleshooting Process Plant Control, by Norman P. Lieberman
Copyright © 2009 John Wiley & Sons, Inc.

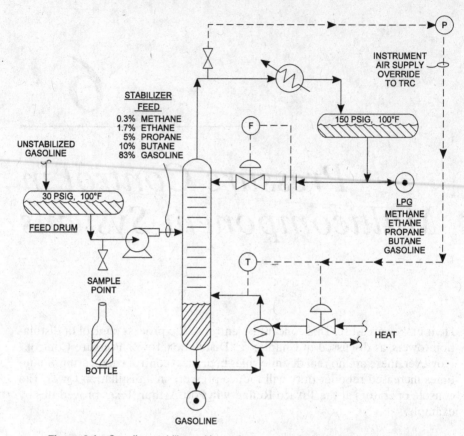

Figure 6-1   *Gasoline stabilizer with condenser capacity limiting pressure control*

the plant flare. The unaccounted-for methane and ethane could only escape from the reflux drum dissolved in the overhead LPG liquid product. Steve is still a good friend and a fine engineer, but this was a fatal error. As the summer progressed in Port Arthur, the operators reported to management that the tower pressure was becoming excessive. Routine flaring of off-gas from the reflux drum was against the law in Texas. As the cooling water temperature flowing to the overhead condenser increased, excessive tower pressure was becoming a serious problem.

HEAT DOES NOT ALWAYS MAKE PRESSURE

Steve tried to reduce the tower bottoms temperature set point so as to reduce the reboiler heat duty. Less reboiler duty would unload the condenser. The heat introduced to the tower must be removed by the condenser. The fact that the tower pressure was increasing as the cooling water temperature increased proved that the stabilizer was limited by condenser capacity. This was correct. But when Steve directed the console operator to reduce the temperature set point (TRC) on the tower reboiler, the tower pressure went up instead of down (see Chapter 5, "Distillation Tower Pressure Control").

Rather than accepting the reality of this situation, Steve became angry. He concluded that the instrument technicians and plant operators had failed to carry out his instructions. He may have referenced my book *Troubleshooting Process Operations* (published by McGraw-Hill; ISBN-0-87814-348-3), where I state that "heat makes pressure."

The panel operators had found that increasing the tower's bottom temperature TRC would lower the tower pressure, even though the tower top temperature would increase. Note, as shown in Figure 6-1, that the reflux rate to the stabilizer, which is on flow control, is assumed to be constant. Increasing the reboiler TRC set point would definitely increase the condenser duty. This ordinarily would increase tower pressure. Why then the apparent contradiction between theory and practice? There are two reasons:

- **Reason Number One**—With the reflux rate constant, the tower top temperature must increase as the bottom TRC is increased. The available condenser capacity (Q) is calculated from the heat transfer equation:

$$Q = U \cdot A \cdot (\text{Delta T})$$

where

- Q = Condenser duty, BTU/hr
- U = Heat transfer coefficient, $\text{BTU/hr/ft}^2/°\text{F}$
- A = Exchanger area, ft^2
- Delta T = the temperature difference between the cooling water and the condensing hydrocarbon vapors, °F

Let us assume that both the area and the heat transfer coefficient are constant. Determining delta T is a complex calculation for condensation of multi-component vapors. Therefore, for ease of explanation let us further assume that delta T will directionally increase as the tower top temperature goes up. This increase in delta T will increase the condenser capacity. This will more or less offset the extra condenser duty required by the extra reboiler duty.

- **Reason Number Two**—With the reflux rate constant, the tower top temperature must increase as the bottom TRC is increased. The higher tower top temperature increases the pounds of gasoline distilled overhead into the LPG product shown in Figure 6-1. The extra gasoline in the overhead product acts as an absorption oil to dissolve the light methane and ethane in the reflux drum.

From the process control perspective the instrument air override signal from the tower top pressure transmitter to the reboiler TRC worked nicely. However, the effect was just the opposite of what Steve had expected. I revised the process control scheme so that when the tower top pressure exceeded its set point the pressure transmitter (PRC) would override the bottom temperature controller (TRC). That is, the heat input to the reboiler would be increased to lower the tower pressure back down to its set point. Steve said this was all counterintuitive. The extra heat, Steve said, should not be reducing the tower pressure.

EFFECT ON LPG QUALITY

The operators were quite content with the revised pressure control override loop. After all, they observed, I had just automated what they had already been doing manually. But the Texaco Refinery management was far from happy. The LPG product now had 5–10% gasoline. And as the cooling water temperature in Port Arthur increased during the summer, so did the gasoline content of the LPG. Downstream of the gasoline stabilizer, the LPG product was refractionated into propane and butane. The butane was sold as a chemical plant feedstock to a nearby cracking plant that produced olefins. This plant did not wish to purchase butane with an erratically high gasoline content. Also, gasoline had a higher product value as finished gasoline than in chemical plant feed.

The management of the Texaco refinery found a solution to this quandary. They fired my friend Steve. The moral of this story is, don't draw unstabilized hydrocarbon samples from 30 psig feed drums into glass bottles. Use a pressurized steel sample bomb. Also, heat does not always make pressure in multicomponent hydrocarbon distillation towers. Finally, process control design requires an understanding of complex vapor-liquid equilibrium issues.

TEST QUESTION

Referring to Figure 6-1, assume the reboiler duty is constant. The operator increases the reflux rate. The tower top temperature and bottom temperature both drop. The condenser duty remains constant because the reboiler duty is kept constant.

The overhead product rate of LPG drops because I've raised the reflux rate. The tower is limited by condenser capacity, just like it was for Steve. Question: What happens to the tower pressure?

The answer is that the pressure goes up. But why does the tower pressure increase? Here are the reasons.

- There is less heavier material in the LPG overhead product to help absorb the lighter components.

- The temperature difference between the cooling water and condensing vapors has been reduced at the lower tower top temperature. This in effect reduces the condenser capacity.

I first noticed this superficially strange response to a higher reflux rate working as a process superintendent for Amoco Oil in Texas City in 1974.

REFRIGERANT COMPOSITION

This is a story from 1976 that I am particularly proud of. You can see how proud I am in the photo in Figure 6-2. I'm the tall young man in the white shirt and tie at the back of my group of operators. My alkylation unit had set a new production record using a new depropanizer reboiler control scheme. The alkylation unit had always been limited by the refrigeration capacity. The refrigerant was needed to keep the reactor, shown in Figure 6-3, below 60 °F. There were two separate issues that limited the refrigerant flow:

- The compressor suction pressure P-1 had to be at least one psig. A lower pressure could potentially draw air into the circulating isobutane refrigerant through a small leak. The air could accumulate in the refrigerant receiver and potentially detonate.
- The compressor discharge pressure at P-2 could not exceed the 100 psig relief valve pressure on the refrigerant receiver vessel.

Figure 6-2   *No. 2 alkylation unit sets record, March, 1975 at 23,851 B/D. Author is in back row with white shirt and tie.*

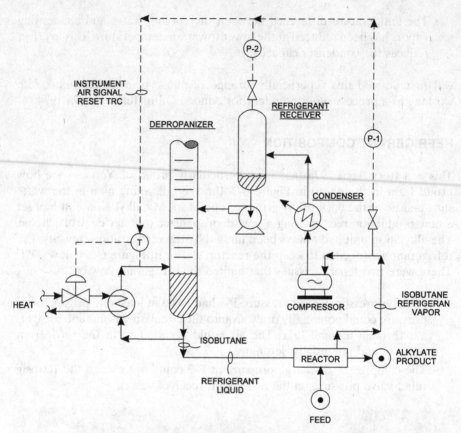

Figure 6-3 *Maximizing refrigeration capacity by override control*

If the pressure at P-1 was too low, the panel operator would slow down the steam turbine-driven refrigeration compressor to raise the compressor suction pressure. The slower speed forced a reduction in the reactor feed rate and thus reduced alkylate production rate.

If the pressure at P-2 was too high, the panel operator would reduce the reactor feed rate. As the reaction was exothermic, this reduced the refrigerant vapor flow and unloaded the condenser. The cooler condenser outlet temperature reduced the isobutane refrigerant pressure in the refrigerant receiver. Again, the reduction in feed reduced alkylate production.

I have called the refrigerant isobutane. This is not accurate. The refrigerant was a multi-component mixture of:

- Isobutane   80–90%
- Normal butane   5–10%
- Propane   5–10%

Excess normal butane and propane were brought in with the alkylation unit reactor feed. Excess normal butane left with the alkylate product. Excess propane was distilled from the refrigerant in the depropanizer shown in Figure 6-2.

ADJUSTING REFRIGERANT COMPOSITION

Let's consider the low compressor suction problem first. This was the bottleneck to making more alkylate product when the cooling water temperature was less than 85 °F. I didn't want my operators to slow down the refrigeration compressor to raise the compressor's suction pressure. So I tried reducing the depropanizer bottoms temperature by 5 °F. This increased the propane content of the refrigerant liquid. Propane at 60 °F has four times the vapor pressure of isobutane. Slowly, the compressor suction pressure increased at a constant reactor temperature of 60 °F. I adjusted the depropanizer bottoms temperature (TRC) to maintain a minimum safe compressor suction pressure of 1 psi.

Let us next consider the high compressor discharge pressure as limited by the 100 psig relief valve setting on the refrigerant receiver vessel. This was the bottleneck to making more alkylate product when the cooling water temperature in Texas City was over 90 °F. I didn't want my operators to reduce the reactor feed to unload the compressor discharge refrigerant condenser. So I tried increasing the depropanizer tower bottom temperature by 5 °F. This reduced the propane content of the refrigerant liquid. Propane at the condenser discharge temperature of 105 °F has three times the vapor pressure of isobutane. Slowly the compressor suction pressure decreased at a constant refrigerant receiver temperature of 105 °F. I adjusted the depropanizer reboiler outlet TRC to maintain the maximum permissible refrigerant receiver pressure as limited by the vessel's relief valve setting.

To summarize, in the cool mornings P-1 would reset by means of override control, the temperature set point controlling the heat input to the depropanizer reboiler. In the hot afternoon, P-2 would reset by override control the temperature set point that controlled heat input into the depropanizer reboiler. I describe how override control works in Chapter 12.

EXPLAINING MULTICOMPONENT REFRIGERATION TO THE PANEL BOARD OPERATORS

In 1974 when I became supervisor of #2 Alky in Texas City the operators rotated through all four operator positions every month. As there were four shifts, each month I had sixteen different console operators. Texans are all strong individualists. Every one of the sixteen operators wanted to run the unit differently. I spent two months explaining and demonstrating the use of the depropanizer bottoms temperature as a tool to optimize refrigerant composition so as to maximize the production of alkylate.

I never find it productive to automate a new control scheme unless the operators understand and agree with the new control method. They will only switch the automatic features off and operate on manual if the control engineer has not secured operator buy in and comprehension.

Thirty-two years have raced by since the photo in this chapter was taken. It's nice to recall how I applied my chemical engineering education to make what I mistakenly thought then was a socially valuable product. We produced over one million gallons of gasoline each day in March, 1976. One thing I learned for sure in Texas City that year. Ice cream and cake is an excellent way to promote a new and novel mode of process control by manipulating the composition of a multicomponent refrigerant.

7

Optimizing Fractionation Efficiency by Temperature Profile

The best process control method to optimize a tower's fractionation efficiency is to use online gas chromatographs. Most process plants do not have many such expensive and high-maintenance intensive on-stream analyzers. However, almost all towers have reliable top and bottom temperature transmitters. To optimize fractionation efficiency in trayed towers, we should consider four process parameters that we should control in an optimum fashion:

- Relative volatility
- Entrainment
- Channeling of vapors and liquids
- Internal reflux rates

The definition of relative volatility is:

$$\frac{VP_L}{VP_H}$$

- VL_L = the vapor pressure of the light component at a particular temperature
- VP_H = the vapor pressure of the heavy component at that same temperature

Troubleshooting Process Plant Control, by Norman P. Lieberman
Copyright © 2009 John Wiley & Sons, Inc.

The larger the relative volatility, the easier it is to fractionate. For example, increasing the relative volatility between propane (the light component) and butane (the heavier component) by 10% would permit one to achieve the same degree of fractionation with 10% less reflux flow. This would also save about 10% of the reboiler duty. Incidentally, we would have to reduce the C_3–C_4 splitter pressure from 320 psig to 180 psig to achieve this 10% benefit.

Unfortunately, the lower tower pressure would increase vapor velocity even at a lower reflux rate by about 50%. This would make the fractionation worse, not better, if the higher vapor velocity promoted entrainment of liquid from the tray below to the tray above. This is called jet flood. But perhaps the larger vapor velocity could be beneficial rather than harmful. Valve trays or sieve trays or grid trays all suffer from potential tray deck weeping and vapor-liquid channeling at reduced vapor velocities. This is called tray deck dumping. Perhaps then, lowering the tower pressure to achieve a greater relative volatility might actually improve fractionation efficiency even further by reducing any vapor-liquid channeling.

As I lack X-ray vision, how can I tell how best to adjust tower operating pressure to maximize fractionation efficiency at a constant reflux rate? My answer is tower delta T. Let's assume a tower is running at 20,000 BSD of top reflux. The temperature difference between the top and bottom of the tower is:

- Top temperature = 130 °F
- Bottom temperature = 210 °F
- Delta T = 80 °F
- Tower pressure = 180 psig

Without changing the reflux rate, I slowly lower the tower pressure from 180 psig to 150 psig. The tower temperature profile is then observed to be:

- Top temperature = 120 °F
- Bottom temperature = 204 °F
- Delta T = 84 °F
- Tower pressure = 150 psig

An increase of 4 °F for the delta T is an indication that fractionation efficiency has improved. But why? Was this due to:

- Less tray deck weeping?

or

- Enhanced relative volatility?

I don't know and I don't really care! What I do know is that I can reduce the tower reflux from 20,000 BSD to 16,000 BSD to drop the delta T back to 80°F. What I do care about is that the reduction in reflux rate has saved 6000 lbs/hr of 100 psig steam worth of reboiler heat input. This is valued at $1200/day or $400,000/year in energy savings (in February 2008 with crude at $100.68 U.S. this morning).

ADVANCED COMPUTER CONTROL

The idea of optimizing tower pressure to save energy is not a new idea. One of my clients has a longstanding practice called the "Pressure Minimization Program." Their computer looks at a tower delta T. Then the computer tries to reduce the tower pressure very slowly on automatic closed-loop computer control. If the tower delta T goes up or remains the same, the computer reduces the tower operating pressure another notch. Note that reducing the tower pressure, even at constant fractionation efficiency and thus a constant tower delta T, has a secondary benefit. The benefit is that less energy is required to heat the feed to the tower bottom's outlet temperature.

If the tower delta T goes down, when the pressure is reduced a notch, the computer increases the pressure bit by bit until the delta T starts going down. My claim that this represents advanced computer control technology is part of the reason I was ejected from the faculty of LSU in Baton Rouge (see Chapter 2). I suppose it's all rather elementary computer control technology. But it does work nicely.

SUPER-ADVANCED COMPUTER CONTROL

Sometimes more reflux improves fractionation. But sometimes more reflux makes fractionation worse. More reflux means more reboiler duty, because the reflux is generated in the reboiler. But more reboiler duty means greater vapor velocities. And greater vapor velocities promotes entrainment. Too much entrainment will make fractionation worse instead of better because of jet flooding. How can one tell if the extra reflux is promoting too much entrainment of the liquid between trays without employing X-ray vision?

Simple! Use the tower delta T. If I increase the reflux rate and the reboiler duty increases proportionately, then the tower delta T will go up or down. If the delta T goes down, then the super-advanced, closed-loop, computer-automated control will reduce the reflux rate a notch. If more reflux increases the delta T then the computer, bit by bit, will increase the reflux rate until the tower delta T stops increasing.

The problem with this closed-loop control is that reflux is expensive. The heat to generate the reflux comes from the reboiler, which is consuming costly 100 psig steam. Thus, the benefits of the increased fractionation efficiency must be weighted against the cost of the incremental 100 psig steam. This extra complication is why I call this Super-Advanced Computer Control.

I seem to recall that my reference to this rather simplistic computer application as "Super-Advanced Control" was one of the primary causes for my ejection from my seat at the graduate school faculty at Northwestern University in Chicago (see Chapter 1).

The above optimization technique assumes that there is sufficient excess condenser capacity to permit a reduction in tower pressure. It also assumes that there is sufficient excess reboiler duty to support an increase in the tower's operating pressure and reflux rate. Also, I have assumed that the tower is not in fully developed flood, as discussed below.

FULL FLOOD

If a column is fully flooded, liquid is being massively forced out of the vapor line. The liquid is cold liquid from the reflux drum. The carryover of massive amounts of cold reflux will precipitatively suppress the tower top temperature, but without any concurrent improvement in fractionation efficiency.

Without X-ray vision how can I know when fully developed flood is occurring? There are three independent alternate methods I can use:

- Climb to the top of the tower and open up the two-inch atmospheric vent and see if liquid gasoline squirts out, a technique not exactly conducive to automatic closed-loop computer control.
- See if an increase in the reflux rate (with the reboiler on automatic temperature control—TRC) causes the cooling water outlet temperature to rise. If not, the tower is suffering from fully developed flood. But the cooling water outlet temperature is not typically transmitted to the operator's control panel.
- Determine whether an increase in the reflux rate with the reboiler on TRC causes the steam flow to the reboiler to increase. If not, the tower is definitely suffering from fully developed flood. This information is accessible from the control room console.

In this case, lowering the tower pressure or raising the reflux rate will make fractionation worse because of flooding, even though the tower's delta T is going up. Override control, as described in Chapter 12, would then be incorporated into your computer control to stop the tower pressure control from reducing the tower pressure any further, or to stop the top reflux valve from opening any further.

I rather hope this last section illustrates why we need general process experience to be a good Process Control Engineer. I asked one young graduate to read this section of my manuscript. He said that I was ignoring the simplest method to identify fully developed flood. That is, delta P across the trays would exponentially increase upon flooding.

Not so! Suppose it was only the top tray that was flooding. A common enough problem due to dirt in the reflux or volatile salt sublimation due to the overhead vapors contacting the cold reflux. Then we would suffer the slings and arrows of outrageous full flood without the symptom of high tower delta P to alert us to our fate. Another example as to why experience counts in our work as control engineers.

THE CASE FOR FEED PREHEAT

Figure 7-1 shows the propane-butane splitter at the Good Hope Refinery in Norco, Louisiana. I only had visitation rights with my children every second Saturday. My ex-wife had custody this weekend. Feeling low, I decided to hang out at the East Plant Control Room. My pal Dee Adams—who also had three children, but from three ex-wives or girlfriends—was working the day shift.

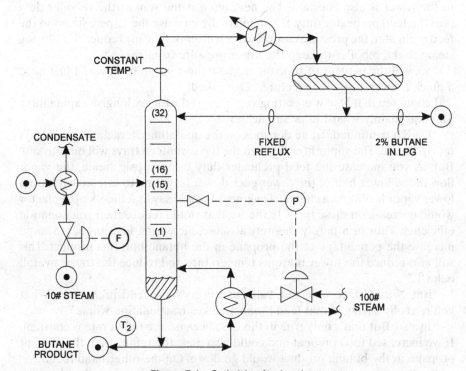

Figure 7-1 Optimizing feed preheat

"Norm, I need to minimize the propane content of my butane bottoms product. My reflux pump is running at maximum. So my reflux rate is constant. Also, I'm keeping the butane in the overhead LPG (propane) produced fixed at 2% C_4's in propane to meet my LPG spec. The heat input to the reboiler is on automatic pressure control (see Chapter 5, "Distillation Tower Pressure Control"). More heat, more pressure. The incremental reboiler duty is holding my tower pressure constant."

"So, Dee, what's the problem?" I asked.

"Well, Norm, for one thing, my ex-girlfriend. You know Marlene in Accounting. Well, she's pregnant."

"No, Dee, what's your problem with the splitter?" I asked. "I can't help about Marlene."

"Yeah, Norm. The splitter's also a big problem. What should I do to the flow of 10 psig steam to the feed preheater to minimize the percent of propane in butane? I don't have time to start in with lab samples. Lisa and I, that cute secretary from Human Resources, are meeting for coffee at six. Should I raise or lower the 10 psig steam to the feed preheater?"

"Look, Dee. Because you've kept the tower pressure constant and the overhead product composition constant, the phase rule tells us the tower top temperature is also constant (see Chapter 3, "Dependent and Independent Variables"). Because you've kept the reflux rate constant, then the heat input to the tower is also constant. The heat input is the sum of the reboiler duty plus the feed preheater duty. If you manually increase the 10 psig steam to the feed preheater, the pressure control valve will automatically reduce the 100 psig steam to the reboiler to keep the tower pressure from rising."

"So, will the percent of propane in the butane go up or down? I just need a quick answer; not a long lecture," Dee asked.

I could see that Dee was getting very annoyed with my longish explanation. This was mainly what I hoped to achieve.

"Dee," I continued. "It all depends on the operating characteristics of trays one to fifteen. The vapor flow through the top seventeen trays will be constant. But, as you increase the feed preheater duty using 10 psig steam, the vapor flow in the lower half of the tower goes down. Trays one to fifteen will have a lower vapor traffic. As a chemical engineer I would say that the stripping factor would decrease on these trays. In theory, that would reduce their fractionation efficiency. Thus in a purely theoretical sense, increasing feed preheat should increase the percentage of the propane in the butane bottoms product. This will also reduce the tower bottoms temperature and reduce the tower overall delta T."

"But Norm," Dee objected, "all our towers have feed preheaters. What you're really saying is that feed preheat makes fractionation worse."

"Just so. But that's only true in this case because the reflux rate is constant. If we increased feed preheat and could also raise the reflux, then the percent propane in the butane product would go down. On the other hand if . . . ."

Really annoyed, Dee Adams suddenly cut the 10 psig steam flow by 5000 pounds per hour. The 100 psig PRC steam valve shown on Figure 7-1, supplying heat to the reboiler, opened. The tower bottoms temperature went down. The colder bottoms temperature was a sure sign that the percent propane in the butane product had gone up.

"Damn it, Norm! I'm in a rush. This worked opposite to what you said. Stripping harder on the bottom trays should have increased the tower delta T. The bottoms temperature at T-2 should have gone up, not down. What's up?"

"Okay, Dee, but you didn't let me finish. Lisa will wait for you. I was saying that if the bottom trays are suffering from entrainment, reducing their vapor load would improve their fractionation efficiency. You need to adjust the feed preheat duty depending on whether the stripping section trays have lost fractionation efficiency due to entrainment," I concluded.

Dee started to change into his street clothes. He really was in a rush. It was still only ten to six. "Unlike you Mr. Lieberman, I don't have X-ray vision. How do I know if the stripping trays are entraining liquid or not?"

"Dee, Lisa's only nineteen. She'll wait for you. Your first wife has a son the same age. Anyway, you don't need X-ray vision. All we need to do (see Fig. 7-2) is adjust the feed preheater duty to maximize the splitter's bottom

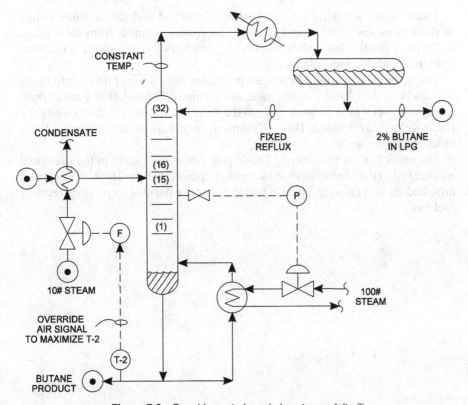

Figure 7-2 Override control maximizes tower delta T

temperature. You don't have to worry about the stripping factor or liquid entrainment. All you have to do is maximize the tower delta T. With our fixed tower top temperature you simply need to adjust the steam flow to the feed preheater so as to maximize T-2, the tower's bottom temperature."

I then configured the 10 psig steam flow control (FRC) to be overridden by the change in the tower's bottom temperature, another advanced application for closed loop automated computer control (as per Chapter 12, "Override and Split-Range Process Control"). This control loop was not intended to hold a fixed tower bottom's set point temperature. The flow of 10 psig steam to the feed preheater was varied so as to continuously maximize T-2. This was consistent with Dee Adams's objective of minimizing the percentage of the light key component, propane, in the butane bottom product, based on the following assumptions:

- Reflux rate fixed
- Overhead product purity constant
- Tower pressure constant
- Feed preheat being the independent variable and the reboiler duty being the dependent variable

I was wrong, though. Dee and Lisa were married and raised a fine family of three sons. Lisa started an extremely successful business from their home in Corpus Christi selling safety awards and company logo tee shirts to the local refineries and chemical plants.

The actual automation of the feed preheater duty was not done until many months later. But I did demonstrate, using manual control, that I could optimize the tower's heat balance and maximize fractionation efficiency using the tower delta T as a guide. This technique is as old as distillation itself. Even older than your author.

The other benefit was energy. The 10 psig steam was excess in the plant and was vented to the atmosphere at the refinery power station. The 100 psig steam flow had to be generated in the boiler house by burning expensive refinery fuel gas.

8

Analyzer Process Control

Captain Hallikanian lived on an expensive yacht in an exclusive marina in San Francisco Bay. During World War II the captain was the director of Under-Sea Warfare Technology for the U. S. Navy. Afterwards he made a fortune inventing analyzers for the refining industry. My visit to his yacht was to purchase his design for an online analyzer for measuring the strength of spent sulfuric acid from an alkylation unit. Captain Hallikanian had also invented on-stream analyzers for:

- End point
- Boiling range
- Density
- Viscosity
- Flash point

At the Good Hope Refinery where I was the plant manager in 1984 we had many other on-stream analyzers:

- Hydrogen sulfide in fuel gas
- Carbon monoxide in furnace flue gas
- Refractive index of gas oil (correlates with H_2 content of oil)
- Sulfur content of diesel oil

Troubleshooting Process Plant Control, by Norman P. Lieberman
Copyright © 2009 John Wiley & Sons, Inc.

- Light hydrocarbon molal concentration in absorber off-gas
- Ethane content of LPG
- Color of gas oil (correlates with asphaltines)
- pH of waste water
- Opacity of fluid cracking unit vent gas
- Oxygen in furnace flue gas

For many laboratory analyses, an online analyzer with a repeatable electronic output can be purchased. "Repeatable" does not mean the output from the analyzer is continuous.

CONTROLLING DIESEL DRAW-OFF RATE

In a refinery crude distillation tower, we wish to maximize diesel oil production and minimize gas oil production. Figure 8-1 shows these two product draw-offs. Diesel oil can be sold as a finished product. Gas oil must still be processed in catalytic cracking units to turn it into gasoline and home heating oil. That is why the diesel oil product is worth 5% more than gas oil.

As we increase the diesel draw-off rate, the composition of the diesel becomes heavier, in the sense that the diesel oil boils at a progressively higher temperature. Let's assume I will vaporize 95% of the diesel in the lab at atmospheric pressure. The temperature at which 95% of the diesel vaporizes is called the 95% point of the diesel product. Typically, if the diesel is used in trucks, the 95% point should not exceed 650 °F. It is this 650 °F boiling point that limits the amount of diesel that can be extracted from the crude distillation column.

As the diesel draw-off rate is increased, the diesel oil draw-off temperature also increases. Historically, panel board operators have used changes in the

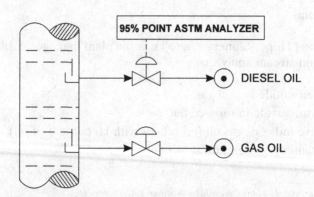

Figure 8-1 *Maximizing diesel production with an online analyzer*

diesel draw-off temperature as a guide to adjust the diesel draw-off rate. But from the Control Engineer's perspective, it's not that simple. The diesel oil 95% point is a function of not only draw-off temperatures but also:

1. Tower pressure
2. The composition of the crude charge
3. The internal reflux on the trays
4. Tray fractionation efficiency
5. The percentage of lighter jet fuel components in the diesel oil
6. The ratio of steam to hydrocarbons at the diesel draw-off tray
7. The overall tower heat balance

In the 1960s, as a young process engineer, I worked for Amoco Oil in Whiting, Indiana. I was assistant technical service engineer on No. 12 pipe still, the world's largest single train crude unit. We had a primitive computer model that advised the panel operator as to how to adjust the crude column diesel draw-off rate to maximize diesel production without exceeding the 95% boiling point specification. This was not a closed-loop control application. In practice, it didn't work. Operating personnel repeatedly proved they could, after decades of hands-on experience, do a better job of maximizing diesel oil production without the help of "Lieberman's computer nonsense." My problem was the difficulty in correcting the optimum diesel draw-off temperature for the above seven variable parameters. For example, my computer model ignored variations in tray efficiency due to alterations in delta P through the tray decks, even though this parameter was known to affect fractionation efficiency between diesel and gas oil products.

Forty-three years have passed, with vast improvements in computer modeling and control. Most of my clients have direct computer control of such parameters as the diesel oil product draw-off rate. Still, it's difficult to mathematically correct for variations in tray efficiency. If for no other reason, tray fractionation efficiency varies with rates of tray deck fouling due to corrosion, which is unknown and unknowable.

DIRECT ANALYZER CONTROL

I recall not too many months ago watching an excellent panel board operator on a crude distillation unit at work in Lithuania. He had a modern 95% boiling point analyzer that displayed its result on a strip chart recorder. The result was updated on the strip chart every 5 minutes. Every 20 minutes or so this gentleman would glance at the strip chart. He would then adjust the diesel draw-off controller. If the 95% point was above 650 °F, he would close the valve by 1%. If the 95% point was below 650 °F, he would open the diesel draw-off valve by 1%. The diesel oil was never at 650 °F and 95%. It just wandered around the target. But it didn't matter. The diesel oil product was all mixed in a down-stream storage vessel. My client's objective of maximizing diesel production was being achieved by this manual control of the draw-off valve.

I happened to come back into the control room late that night. I had left my reading glasses behind. The operators on the night shift did not know me. I spoke no Lithuanian, they spoke no English. I watched the operators for an hour, during which no adjustments were made to the diesel draw-off rate. The operator did not check the online 95% point analyzer, even though it had drifted down to 620 °F. When I tapped on the strip chart to draw his attention to the low 95% point, he opened the diesel draw-off valve by 10%. Rather soon, the 95% point jumped to 690 °F, which is extremely high and would negatively impact the downstream hydro-desulfurizer catalyst.

The solution to this problem was to duplicate, via automatic closed-loop analyzer control, what I had seen the excellent operator on the day shift do manually. The result was quite satisfactory. Of course, analyzers are expensive and require continuous maintenance by qualified craft personnel. Mathematical modeling is comparatively inexpensive, and no craft maintenance is needed. Knowing the complexities of creating a representative computer model for a distillation unit, I would prefer direct analyzer control of critical process variables such as the diesel draw rate. This gets back to one of Dr. Shinsky's rules, "If you can run it on manual, then I can make it work in automatic. If it won't work in manual, it can't be automated."

DE-ETHANIZER REBOILER CONTROL

In 1989 I retrofitted a de-ethanizer at the Chevron Salt Lake City Refinery. The tower had a tendency to oscillate between excessive ethane in the bottoms and flooding. The reboiler heat input was on reboiler outlet temperature control. When the temperature was above its set point, the steam flow to the reboiler would cut back and excessive ethane would dump into the bottoms because of reduced vapor traffic in the stripping trays (see Fig. 8-2). When the reboiler outlet temperature was below its set point, steam flow to the reboiler would increase. The surge of vapor through the stripping trays would cause tray flooding. Flooding was indicated by the carryover of bottoms product into the de-ethanizer overhead off-gas.

I solved this problem with a design change to the reboiler control as shown in Figure 8-2. A new online gas chromatograph was installed. The analyzer measured percentage of ethane in the bottoms every 5 minutes. If the ethane was above the target, then the steam control valve would open by a single percent. If the ethane was below target, the valve would close by 1% of valve position. This slow, but steady, control method produced a propane product that continuously had somewhat too much ethane, or somewhat too little ethane. However, as the propane was stored in a tank with a week's worth of production, it averaged out to the 2% ethane spec for refinery-grade LPG.

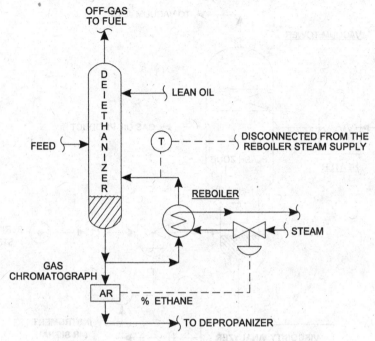

Figure 8-2   Heat input directly controlled by product spec

And, most importantly, the carryover of lean oil into the refinery's fuel gas system was eliminated.

CONTROL OF ASPHALT VISCOSITY

I was working for a small 9000 BSD asphalt refinery in San Francisco. This plant produced a dozen different paving asphalt products. Each had its own viscosity specification. The asphalt was produced from the bottom of a vacuum distillation column, as shown in Figure 8-3. There were three parameters that the operators used to control the viscosity of the paving asphalt:

1. Tower top vacuum
2. Heater outlet temperature
3. Bottoms stripping steam rate

The tower top vacuum was an awkward adjustment. It involved a local manipulation of the motive steam valve to the primary vacuum steam jet. The heater outlet temperature adjustment was easy to make, as the control was a panel board instrument normally run in auto. However, changes in this

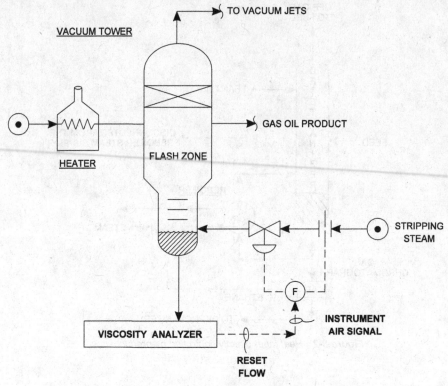

Figure 8-3 *Control stripping steam for asphalt viscosity specifications*

variable affected the required amount of combustion air. The combustion air was controlled by moving eight burner registers in the field, underneath the heater. In practice, frequent changes in the heater outlet temperature resulted in the outside operators leaving all eight air registers 100% open. This caused a reduction in heater fuel efficiency due to the high oxygen content of the heater flue gas.

The best option was changing the stripping steam rate at the bottom of the vacuum column. This control modification had the advantages of:

- The control valve was already on flow control on the panel board.
- The effect on asphalt viscosity was very quick, that is, within 5 minutes.
- There were no effects on the other areas of the process caused by altering the bottom stripping steam rate.

There was an existing continuous online viscosity meter. I spent a day experimenting with changing the stripping steam rate and observing the effect on asphalt viscosity. As shown in Figure 8-3, I then integrated the viscosity meter into a closed-loop automatic control to the stripping steam flow control valve. As before, valve position was changed by fixed increments, depending on whether the paving asphalt viscosity was above or below the target specification for the particular grade of asphalt being produced.

One of the side benefits of this automated control of the product viscosity was that the four operating shifts all started to use the same method to control viscosity. Previously, some shifts had used temperature or vacuum or stripping steam to meet the viscosity specifications. Now, all the operators relied on variations of the bottoms stripping steam rate to make the appropriate grades of paving asphalt.

9

Fired Heater Combustion Air Control

My mother was the smartest person who ever lived. One of her notable insights was "Trouble makes trouble." This was her way of warning me about the dangers of positive feedback loops. Enrico Fermi and Leo Szilard, the parents of the atomic bomb, also knew about positive feedback loops. They proved to the world that neutrons make neutrons.

Panel board operators are familiar with the hazards of positive feedback in combustion equipment. They call the problem:

- Stalling out the heater
- Smothering the heater with fuel
- Bogging down the heater
- Burning too rich
- Starving the box for air

These terms all mean that the ratio of combustion air to fuel is less than optimum. What, though, is meant by the optimum combustion air rate? How can we control a fired heater to reach this optimum air flow?

The optimum air flow is definitely not that required to achieve complete combustion. There is no such thing as complete combustion. I know this for sure. As the air flow to a furnace increases, the carbon monoxide (CO) decreases. However, regardless of the amount of excess air used, there will always be a residual amount of CO in the furnace flue gas. There are two ways

---

Troubleshooting Process Plant Control, by Norman P. Lieberman
Copyright © 2009 John Wiley & Sons, Inc.

to reduce CO and other partially oxidized hydrocarbons in the heater flue gas:

1. More combustion air
2. Improved mixing between the fuel and the combustion air

I have often tried to use more air, and this has reduced CO in the flue gas, but never to zero. I have tried to improve air fuel mixing efficiency. This has also reduced CO in the flue gas, but not to zero. The only way to reduce partially oxidized hydrocarbons such as CO in the flue gas to zero is to have perfect air fuel mixing efficiency. As there is no such thing as perfect mixing, there can be no such thing as complete combustion. If perfect mixing of air and fuel was possible, then using the stoichiometric ratio of the reactants would oxidize all the carbon to CO_2 and all the hydrogen to water. But that is impossible.

CONTROL OF AIR WITH O_2 ANALYZER

Many of my clients have asked me for the best method to use an oxygen analyzer to control combustion air flow. To answer this question let us examine Figure 9-1. Fuel gas flow is constant. Air flow is decreased by closing valve "A." To start with, let's assume valve "A" is 100% open. The analyzer is showing

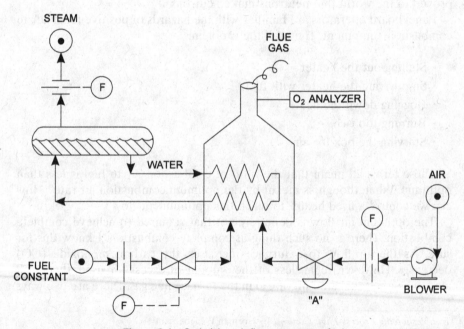

Figure 9-1 *Optimizing air flow at constant fuel*

15% O_2 in the flue gas. By any criteria the combustion air flow is too high. A typical O_2 target is 2% to 3%. The operator therefore begins to slowly close valve "A." Figure 9-2 summarizes the result of the gradual reduction in air flow.

At first, steam production increases because of less heat loss up the furnace stack. As the combustion air flow is reduced, the flow of flue gas is also reduced. It's the flue gas that conveys heat up the stack. If more heat is lost up the stack, there is less heat left over to generate steam. As the operator reduces the air flow, the steam production peaks at 6% O_2. I call this 6% O_2, the point of absolute combustion. The operator now further reduces the air flow trying to reach the 3% O_2 target. But, as noted from Figure 9-2, steam production starts to drop. Below the point of absolute combustion, the concentration of partially oxidized hydrocarbons in the flue gas will increase. These partly oxidized hydrocarbons are in the form of:

- Aldehydes
- Ketones
- Carbon monoxide
- Light alcohols

that is, environmentally objectionable components. The heat of combustion of a partially oxidized hydrocarbon is less than a hydrocarbon completely oxidized to CO_2 and H_2O. Therefore, below the point of absolute combustion steam generation declines. Again, above the point of absolute combustion all of these partially oxidized hydrocarbons would still be found in the flue gas, but in lower concentrations.

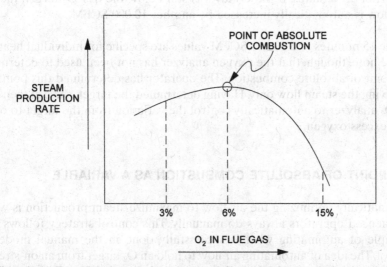

Figure 9-2 *The point of absolute combustion defines the optimum combustion air rate*

The point of absolute combustion is then defined as that combustion air rate that maximizes steam production for the boiler shown in Figure 9-1. This boiler is operating within the following narrowly defined set of parameters.

- Fuel rate is constant.
- The heating value of the fuel is constant.
- Ambient conditions are constant.
- There are no tramp air leaks in the firebox.
- Only steam at a constant pressure and temperature is being generated.

SETTING A TARGET

Clearly one could find the point of absolute combustion by trial and error. The operator tries a variety of combustion air rates until she finds the point at which steam flow is maximized. That is 6% O_2. To reduce air flow below 6% O_2 in the flue gas would be senseless. Steam production would fall and environmental emissions would rise.

Having determined the target of 6% O_2, the operator could now use the oxygen analyzer as a tool for closed-loop automatic control of the combustion air rate through valve "A" shown in Figure 9-1. I discuss in detail in Chapter 8 how this could be done. To use the O_2 for automatic control:

- If the O_2 level is below 6%, the air flow is increased by 10,000 SCFM (standard cubic feet per minute).
- After 15 minutes, if the O_2 level is still below the 6% O_2 target, the air flow is automatically increased by another 10,000 SCFM.

The 15 minutes and 10,000 SCFM values are specific for individual heaters. Please note, though, that the oxygen analyzer has not been used to determine the point of absolute combustion. The operator has determined this point by observing the steam flow only. Having determined the target, she can then use the O_2 analyzer to automatically control the variation from the target to optimize excess oxygen.

THE POINT OF ABSOLUTE COMBUSTION AS A VARIABLE

Automatically optimizing the air flow to maximize steam production is what experienced operators always do manually. This control strategy follows the principle of automating what is successfully done in the manual mode of control. The idea of automating air flow to hold an O_2 target from an on-stream analyzer output is standard in the process industry and forms the basis for many automated fired heater control systems. Unfortunately, in practice, this strategy has a serious drawback.

The problem is that the point of absolute combustion is a variable. It depends upon the air fuel mixing efficiency of the burner. If an operator accidentally leaves a furnace site port open, air is drawn into the firebox. Air that enters a firebox through such an opening, rather than through the burner, cannot mix efficiently with the fuel. The degraded air fuel mixing efficiency increases the excess air required to reach the point of absolute combustion. Other factors that commonly effect air fuel mixing efficiency are:

- Low NO_x burners
- Leaking secondary air registers on burners that are out of service
- Composition changes of the fuel affecting its heating valve
- Air leaks in the firebox
- Burner tip plugging
- Wind direction
- Air temperature
- Changes in the burner heat release rate

As the optimum ratio of air to fuel varies in inverse proportion to mixing efficiency, how can the panel board operator arbitrarily select an O_2 control target of 6%, or 3%, or any value? The problem with automatic closed-loop combustion air control using an oxygen analyzer lies not with the analyzer itself but with the selection of a target. Suppose the oxygen target in the flue gas is lower than the point of absolute combustion. Looking at Figure 9-2, less than the maximum amount of steam is produced. Energy is wasted in the form of partly oxidized hydrocarbons escaping up the stack with the flue gas.

Also, smog producing agents are generated in increasing concentrations. This, while unfortunate, is also a reversible problem. The panel operator just selects a greater O_2 target for the analyzer set point. But there are other potential problems, of an irreversible nature, associated with selecting an excess O_2 target below the point of absolute combustion.

AUTOMATIC TEMPERATURE CONTROL

Most of the process heaters in petroleum refineries and in petrochemical plants are not run on manual control of the fuel as shown in Figure 9-1. They are operated on automatic temperature control of the fuel gas, as illustrated in Figure 9-3. In this sketch the combustion air flow is being reset in the manner described just above by the oxygen analyzer sampling the flue gas. Let's assume that at time zero all is well. The point of absolute combustion is currently 3%, as determined by a trial-and-error experiment done last week.

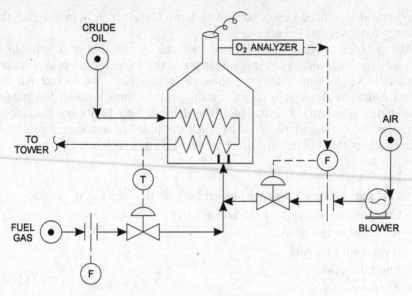

Figure 9-3 *Analyzer sets air flow. This is a bad design*

The current set point is 4% oxygen, which directly and automatically controls the discharge flow from the air blower.

At this time burners start to plug with sulfur deposits. The mixing of air and fuel becomes less efficient. The point of absolute combustion increases from 3% to 5%. This happens without the panel board operator's knowledge. What will next transpire automatically?

- **Step One**—More partially oxidized hydrocarbons will escape up the stack with the flue gas.
- **Step Two**—Heat release per pound of fuel gas will decline.
- **Step Three**—The heater crude oil outlet temperature will begin to fall.
- **Step Four**—The fuel gas regulator will begin to open. However, as the problem was reduced burner efficiency due to plugging with sulfur deposits, the higher fuel gas rate further reduces air fuel mixing efficiency due to excessively high burner tip velocity.
- **Step Five**—The point of absolute combustion now increases from 5% to 6%. The concentration of partially oxidized hydrocarbons in the flue gas escalates.
- **Step Six**—Heat release per pound of fuel gas will decline even further.
- **Step Seven**—The heater outlet temperature will fall further.
- **Step Eight**—The fuel gas regulator will open further, which further reduces the burner efficiency.

The problem is feeding upon itself. This is called a "positive feedback loop." Let's assume that the operator suddenly realizes that he has entered the region of positive feedback and is operating below the point of absolute combustion. He then takes the air flow valve, which is on automatic O_2 analyzer flow control, and resets it from the 4% target to 7% oxygen in the stack. He is now in trouble:

- **Step One**—The combustion air flow suddenly increases.
- **Step Two**—The heater firebox saturated with CO and light alcohol vapors, overpressures because of the sudden release of heat.

The heater may undergo structural damage. The sudden overpressuring of a heater is a common phenomenon in refineries. It is called heater "puffing" or "thumping." When the puffs get big enough, structural damage will result.

COMBUSTIBLE ANALYZER CONTROL

If we had a CO or combustible analyzer instead of an oxygen analyzer in the stack, the problem would have been mitigated. In this case the ratio of air to fuel would have automatically increased, as the burner mixing efficiency was adversely impacted because of sulfur deposits in the burner tip. But I am far from recommending closed-loop analyzer control with such a complex instrument as an online combustible flue gas analyzer.

In the real world analyzers—oxygen or combustible—have a poor online performance record. Therefore, in practice, they are rarely used for closed-loop control. In reality, outside operators will make occasional adjustments to air flow with the furnace stack damper and/or the air registers around the burners. That is, most of the time air flow to a heater is constant.

Let's say a heater is running at the point of absolute combustion. Fuel is controlled automatically to maintain the outlet temperature set point of 700 °F. Suddenly, the feed temperature drops and the heater outlet drops below its set point to 690 °F. What will then transpire?

- **Step One**—The fuel gas regulator control valve will open further.
- **Step Two**—The flow of fuel gas into the firebox will increase.
- **Step Three**—Since we were already operating at the point of absolute combustion, the incremental fuel will not burn.
- **Step Four**—Since the fuel gas is cold and it does not burn, the firebox temperature drops from 690 °F to 680 °F.
- **Step Five**—The fuel gas control valve opens even further. But the incremental fuel just makes the problem worse.

Trouble makes trouble. Let us assume that the operators suddenly realize what has happened. If they realize that they are stalling out the firebox, or smothering the furnace with fuel, or operating below the point of absolute combustion, they may take inappropriate corrective action, meaning that they spin open the secondary air registers around the burners and open the heater's stack damper. The heater may now explode!

The problem is, adding more fuel without more air to a heater running close to the point of absolute combustion creates a positive feedback loop that results in a potentially explosive mixture accumulating in the firebox. The sudden introduction of more air causes the mixture to explode.

A malfunctioning oxygen or combustible analyzer will lead to the same sort of potentially fatal circumstances. All this is quite unnecessary. The correct way to control the flow of combustion air to a heater need not and should not rely on an analyzer.

CORRECT CONTROL OF COMBUSTION AIR

I've had to spend most of this chapter explaining the wrong way to control air flow to a heater. There is too much emphasis in the industry on the use of analyzers. Likely, control engineers who lack experience with real problems are treated to free lunches by analyzer salesmen. The correct and optimum way to control combustion air flow does not require new instrumentation. It does require some elementary use of computer technology.

What is it that we are really trying to achieve? What is our real objective? It's not to reach an arbitrary target for oxygen in the flue gas. It is not to eliminate CO in the stack. What we are trying to achieve is the minimum fuel gas consumption to reach a target heater outlet temperature for a particular feed rate. To reach this objective we should use the control strategy shown in Figure 9-4. The idea of this sketch is to use a feedback loop between the fuel gas flow and the combustion air flow to minimize fuel consumption.

The definition of the point of absolute combustion, when firing an automatic heater outlet temperature control, is the combustion air rate that minimizes fuel consumption. Hence, we want to set up our feedback loop to reach the point of absolute combustion. Air flows above or below this point will increase fuel consumption.

The parameters that are used to calculate the combustion air rate by the online computer are:

- Process fluid inlet temperature
- Process fluid outlet temperature
- Process fluid flow rate
- Process fluid percent vaporization based on product yields
- Furnace efficiency

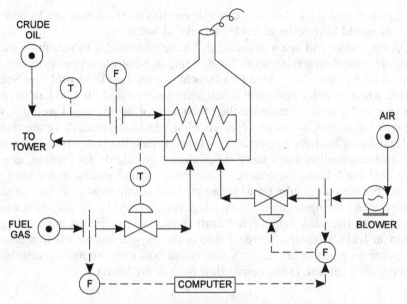

Figure 9-4 *Correct strategy to control combustion air*

- Fuel gas rate
- Fuel gas composition
- Air blower discharge temperature

The furnace efficiency depends on air fuel mixing efficiency and is quite variable. The composition of fuel gas varies over large ranges in a refinery and is quite variable. However, its composition is only reported by the lab several times a day. The process fluid vaporization is also an approximated variable. Thus the calculated combustion air flow is an engineering estimate of the optimum air flow needed.

If the air flow is increased, based on this calculated engineering method, one of two things will happen:

- The fuel gas rate will go up, indicating that the heater is above the point of absolute combustion. The combustion air flow will be reduced by 10,000 SCFM and the reduction reiterated every 15 minutes.

OR

- The fuel gas rate will go down, indicating that the heater is below the point of absolute combustion. The combustion flow will be increased by 10,000 SCFM and the increase reiterated every 15 minutes.

The rate and time intervals noted above must be determined individually for specific units. Natural draft heaters that lack an air blower or automated air registers and stack dampers can still use this method. But the 15-minute interval would have to be extended to several hours.

As the reader will again observe, all I have proposed is to automate using computer and engineering tools for fine-tuning what plant operators always do manually to save fuel. Once I was teaching a seminar in Mossel Bay, South Africa, when an older and quite intelligent operator said, "If I had an oxygen analyzer in the stack to measure the O_2 content of the flue gas, I would never use it to adjust air flow for two reasons. First, Mr. Lieberman, there are tramp air leaks in my heater's convective box that increase the O_2 content of the flue gas after combustion gases leave the firebox. Second, Mr. Lieberman, as you have explained, furnace efficiency varies with air fuel mixing in my burners. As my burners are fouling and being cleaned continuously, air fuel mixing efficiency and the optimum O_2 in the stack ranges from 3% to 7%. Also, every time I move the stack damper, the draft changes in my firebox and so do my tramp air leaks. Actually, the real value of the oxygen analyzer is to measure the tramp air leaks in the heater's convective box, to guard against afterburn or secondary ignition in the convection zone of my heater."

Afterburn

I mentioned before that the consequences of keeping the combustion air rate below the point of absolute combustion were sometimes irreversible. The comments by the operator from Mossel Bay were meant to convey his concerns about secondary ignition or afterburn in the convective section.

The convective section is the bank of tubes stacked in a box, situated atop the firebox. Only hot flue gas, but never fire, should enter the convective box. The tubes and tube supports in the convective box are not thermally rated to be exposed to direct radiant heat transfer from flames. However, if one has located the O_2 analyzer in the stack and the O_2 content of the flue gas has increased because of tramp air leaks, then the indicated flue gas O_2 concentration will be too high. The oxygen in the flue gas will be higher than the oxygen level in the combustion zone because of the tramp air leaks in the convective section. If the firebox is now starved for air, unburned hydrocarbons will migrate up into the convective box. The tramp air leaks will reignite the flue gas. Radiant heat will cause the carbon steel convective section tubes to sag. The tube's fins will oxidize, and the supports will be damaged. An oxygen analyzer in the stack can be used as an indication of tramp air leaks, but should not be used either for firing control or as an adjustment for the combustion air flow.

If you are determined to ignore my advice and use oxygen and combustible flue gas analyzers for control of excess air, then at least locate the analyzers in the firebox just below the lower row of the roof tubes or the bottom of the convective tube bank. But if you do so, remember what my mother said, "Trouble makes trouble." And positive feedback loops definitely are trouble.

10

Sizing Process Control Valves

I am quite an anticommunist. I never disliked communism until I worked in Lithuania, a former Soviet Republic. Figure 10-1 is a sketch of a typical control valve installation in the Mazaikai Naphtha Refinery. Typically, if the process lines are 4 inches, then the control valve ought to be at least 3 inches. Why is this control valve only 1 inch, or about 10% (on a cross-sectional area basis) of its normal minimum size?

In communist countries, engineers were careful to avoid undersizing equipment. Nobody cared about equipment costs. However, if process pumps or piping were too small, the guilty engineer would be accused of industrial sabotage and sent to Siberia. Hence, Soviet engineers vastly oversized motor drivers, turbines, and especially piping. Heat exchanges, furnaces, and coolers were designed for almost zero delta P. The result of this insanity was that 90% of the pressure head developed by the pumps and compressors had to be parasitically sacrificed across control valves. For the control valve to operate in the linear portion of its range, the control valve had to be very small.

SIZING CONTROL VALVES

Calculating the size and trim of a control valve is easily done by a computer. The Process Control Engineer enters the following data:

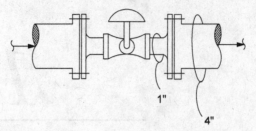

Figure 10-1 A poorly designed control valve installation

- Flow
- Density
- Viscosity
- Delta P at the indicated flow
- Valve position at the indicated flow

The control valve position at normal design flow ought to be 50%. Let's assume that at normal conditions:

- Flow = 100 ft³ per hour
- Density = 50 lbs per ft³
- Viscosity = 4 centistokes
- Delta P to be determined below
- Valve position assumed at 50%

The parameter that will determine the size of the control valve in this example is the delta P we input into the computer program that is sizing the control valve. The larger the assumed delta P, the smaller the control valve. Small control valves with big delta Ps require big pumps driven by big motors, which waste lots of electric power. What is the correct method for the Process Control Engineer to use to specify the delta P for the above conditions? This procedure is:

- **Step One**—Based on discussions with management, establish the maximum reasonable flow that can be anticipated within the next few years. Let us assume projected maximum operating conditions are:

  Flow = 120 ft³ per hour

  Density = 55 lbs per ft³

  Viscosity = 6 centistokes

  Delta P = 10 psi

  Valve position assumed = 90% open

At the anticipated maximum flow I have assumed that the control valve is almost wide open. Also, the valve pressure drop is a small but a reasonable 10 psi. Opening the valve from 90% to 100% will not significantly increase the flow or reduce the delta P.

- **Step Two**—Calculate the pressure drop at the normal design flow of 100 ft^3 per hour. Include the following pressure losses:

  Piping

  Heat exchangers

  Vessels

  Fired heaters

  Filters

  Air coolers, etc.

Do not include the pressure loss through the control valve. Also exclude the extra head pressure required to lift the fluid to a higher elevation and to pump the fluid into a higher operating pressure. Let us assume the above pressure losses (piping, exchangers, vessels, heaters, coolers, etc.) are 100 psi.

- **Step Three**—Note that pressure drop varies with:

$$(\text{Density}) \times (\text{Velocity})^2$$

As the density has increased by 10%, and the velocity has increased by 20%, the pressure drop through the process equipment at 120 ft^3 per hour would increase from 100 psi to:

$$(100\,\text{psi})(120/100)^2\,(55/50) = 158\,\text{psi}$$

I have ignored the increase in viscosity from 4 to 6 centistokes. As long as viscosities are below 8–10 centistokes, I consider changes in viscosity too small to effect hydraulic calculations. At viscosities above 40–50 centistokes, changes in viscosity become critical in pressure drop calculations.

- **Step Four**—The required delta P through the control valve in normal operations is a sum of two numbers:

  The allowable 10 psi assumed when the control valve was 90% open plus the extra 58 psi (i.e., 158 psi minus 100 psi) needed to overcome increased pressure drop, due to frictional losses in the process equipment at the maximum operating conditions.

$$10\,\text{psi} + 58\,\text{psi} = 68\,\text{psi delta P}$$

This is the delta P that will be used to size the control valve for the normal flow of $100\,\text{ft}^3$ per hour at the 50% valve position. But the control valve size must also be checked for a second criteria, that is, the $120\,\text{ft}^3$ per hour flow case when the control valve is 90% open and the delta P is 10 psi. The case that requires the larger control valve is the controlling case for valve sizing.

UNDERSIZING CONTROL VALVES

If a control valve is too small, it will have an excessive delta P when 100% open. By excessive I mean 20% of the pressure developed by the charge pump. In this case I, and everyone else, will partly open the control valve bypass gate valve. From the control aspect and from the safety aspect of process operations, this is bad. Opening a bypass in the field around a control valve partly defeats the purpose of having central control. For instance, suppose an emergency arises and the panel operator has to close the fuel gas control valve to a fired heater. And the fuel gas regulator bypass valve is open! Then what?

Well, he could call the outside operator to shut the bypass around the fuel gas regulator—provided he remembers that it was opened last month. You can imagine how I've become so smart on this subject.

WHY IGNORE CHANGES IN ELEVATION AND OPERATING PRESSURE?

In the preceding calculations used to determine the control valve delta P, we neglected the extra pressure losses (or gains) due to changes in elevation. The reason for this is that changes in pressure due to elevation are constant regardless of flow rates. To expand this concept, we should also ignore changes in pressure due to increases in operating pressure between vessels. For example, if we pump diesel oil from a feed drum of 100 psig to a reactor of 2000 psig, the extra 1900 psig of pump discharge pressure does not affect the control valve delta P calculations. The extra 1900 psig is the same, regardless of variations in the flow rate.

ENERGY LOSSES IN CONTROL VALVES

As the pressure drop across a control valve increases, the horsepower of the motor driving the pump also increases. To obtain an approximate idea of wasted energy expressed in horsepower:

$$HP = \frac{(GPM)(\text{delta P})(SG)}{1200}$$

where:

- HP = Energy lost through control valve in horsepower
- GPM = U.S. gallons per minute at flowing conditions
- Delta P = Control valve pressure drop in psi
- SG = specific gravity (i.e., water is 1.00) at flowing conditions

The 1200 factor includes typical efficiencies for the electric motor and for a mediumsize centrifugal pump.

CHECKING CONTROL VALVE CAPACITY

One of the common problems encountered by the Process Control Engineer is to determine whether an existing control valve is undersized for its service. One possible problem is that the valve may be defective. Or perhaps the internal trim inside the valve is smaller than the engineering records indicate. Basically the question can be stated as, "Is the lack of flow due to a pump capacity issue or a restrictive control valve?"

As there is no theoretical answer to such a question, one should proceed as follows:

- **Step One**—Ask the panel board operator to open the valve to 100% on the control panel. Then observe if the valve is fully open in the field. You will see a valve position indicator, as shown in Figure 10-2. In this sketch the valve is shown 75% open. The valve stem marker moves up and down with the valve stem. The valve position indicator is stationary.
- **Step Two**—Presuming the valve opens 100%, open the control valve bypass valve 3 or 4 turns for a few seconds.
- **Step Three**—If the flow increases by 10%–20% or more, the problem is that the control valve itself, or the control valve trim, is undersized. If the flow barely increases, the problem is that the pump itself is the flow bottleneck. Perhaps a bigger impeller is possible?

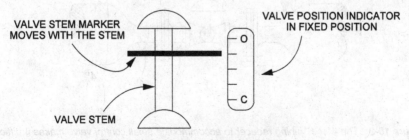

Figure 10-2   Valve position shown 75% open

INCREASING CONTROL VALVE SIZE

To save electricity in the Mazaikai Naphtha Refinery in Lithuania, I embarked on a two-step program:

- Reduce the size of the centrifugal pump impellers so that the pumps produce less discharge pressure.
- Increase the sizes of the control valves.

This program was an utter failure. The refinery maintenance supervisor said my plan was too expensive and too complex. What mistake had I made? Consult Figure 10-1 and Figure 10-3. If the piping for the 1-inch control valve is constructed as in Figure 10-1, increasing the control valve size to 3 inches requires only bolting up the new valve. If the piping is constructed with 4 inch to 1 inch reducers, as shown in Figure 10-3, there is a big mechanical problem. These 4 inch to 1 inch reducers have to be cut out. Next, dual 4 inch to 3 inch reducers must be welded onto the cutoff 4-inch piping.

This is not a job that the pipefitters would easily take on.

I still feel bad when I recall this failed project. It's a fine illustration, though, of a Process Control Engineering problem in a real-world situation.

EFFECT OF OVERSIZING CONTROL VALVES

Sometimes a control valve is too big for the large available delta P and the small flow. Then the control valve will operate in a mostly closed position. This is also bad for two reasons:

- Valve control characteristics will be poor, as in a mostly closed position the valve is operating in a nonlinear portion of its range, meaning that a small movement of the control valve will result in a large change in flow and the flow will be erratic.

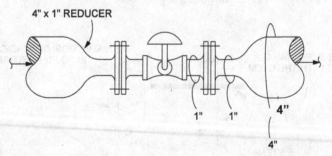

Figure 10-3   *The 4" × 1" piping reducer to accommodate small control valve makes it difficult to increase valve size*

- Second, the high velocities inside the valve will cause erosion and eventual failure of the valve internal trim components.

To return such a control valve to its linear part of its operating range—typically over 20% open—outside operating personnel will partly close an isolation gate valve. It doesn't matter if this is done upstream or downstream of the mostly closed control valve. Most often I see this done at the discharge of the upstream centrifugal pump or at the block valve just ahead of the control valve. This is also bad for two reasons:

- It partially defeats the purpose of central process control. The panel operator no longer has the ability to fully manipulate the flow from his control panel. If the panel operator suddenly requires the full flow, he has to contact the outside operator to open the partly closed isolation gate valve, provided that either operator remembers which gate valve has been partly closed.
- Second, the high velocity between the gate and the seat of the partly closed isolation valve will cause erosion to both the gate and valve seat. Then, when the valve must be shut to isolate a piece of process equipment for repair, the gate valve will leak. That is, its intended function as an isolation valve is destroyed.

Sometimes the best solution to correcting the problem of an oversized control valve is to reduce the size of the impeller in the upstream motor-driven centrifugal charge pump. This is a simple, quick, and inexpensive mechanical change to the pump. If the charge pump is a variable-speed, steam turbine-driven pump, simply reduce the turbine speed by 100 rpm. This is a minor field adjustment to the motive steam governor speed controller set point.

Perhaps we are dealing with a control valve in vapor service on the discharge of a motor-driven fixed-speed gas compressor? Then the number of wheels on the rotor has to be reduced. This is a complex, lengthy, and expensive mechanical change to the compressor. If the gas compressor is a variable-speed, turbine-driven machine, reduce the speed by a few percent. Again, this is a minor adjustment, but unlike for steam turbines, the speed set point change is usually made from the control room console panel rather than in the field.

11

Control Valve Position on Instrument Air Failure

One of the critical decisions that the Process Control Engineer must make is which way a control valve will move upon loss of instrument air pressure. When creating the P&IDs (process & instrumentation diagrams) for a plant, the control valves should be labeled as follows:

- **APO**—Instrument air pressure failure will cause the control valve to close. (APO means air pressure opens the valve.)
- **APC**—Instrument air pressure failure will cause the control valve to open. (APC means air pressure closes the valve.)

The final position of the control valve when there is a total loss of instrument air pressure is a vital safety consideration. For example, in the event of a failure of the instrument air supply, the fuel gas regulator valves to a fired heater must fail in a closed position (APO).

Figure 11-1 shows a control valve arranged within an instrument air supply that opens the valve. A length of ¼-inch copper tubing supplies air pressure on the underside of the diaphragm. There are some rare exceptions to this rule relating the valve action to the air tubing location. But 95+% of the time, Figure 11-1 does apply. The same control valve can be connected with the air supply above or below the diaphragm, so that it can fail in either position.

To change a control valve from APO to APC is a simple procedure. If you feel the current situation is unsafe, mechanically it is simple to change the

Troubleshooting Process Plant Control, by Norman P. Lieberman
Copyright © 2009 John Wiley & Sons, Inc.

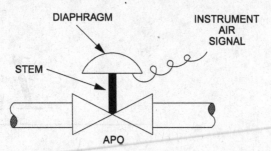

Figure 11-1   *A control valve that will fail closed on loss on of instrument air*

action of a control valve. But this must involve a formal HAZOP review with plant management and operating personnel.

Nitrogen Backup

Of course, it would be best not to lose instrument air pressure in the first place. I accomplished this objective at one refinery by connecting the instrument air supply to the plant nitrogen system. Note that plant air is not a suitable backup for instrument air, as the plant air has not been chemically dried.

There is a potential hazard with the use of nitrogen as a temporary replacement for instrument air. Pneumatic panel board instruments use several cubic feet per minute of instrument air. It is entirely possible to displace air from a control room with nitrogen when the instrument air compressor trips off. Lack of oxygen does not cause any breathing discomfort but does cause death without prior notice. The manager of the refinery observed that this was not one of my better ideas.

Levels

The vessel bottoms level control valve shown in Figure 11-2 will fail in a closed position. If it were to fail in an open position, high-pressure fuel gas would blow through into the storage tank. On the other hand, when this valve fails closed, and the feed to the vessel may continue, the vessel would fill with liquid. Then the liquid would be carried overhead into the fuel gas system. This is also undesirable, but not as bad as fuel gas possibly overpressuring the storage tank.

CONTROL VALVE ON DISCHARGE OF PUMP

Let us assume that we have an inline booster centrifugal pump. On the discharge of this pump there is a control valve. Should we specify APO or APC?

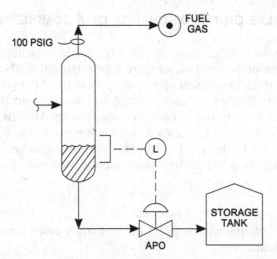

Figure 11-2 Level control valve fails safely in a closed position

In the event of a loss of instrument air pressure, the APC valve would open. The pump discharge flow would increase until the pump suction pressure became too low.

This would cause the pump to cavitate, and the resulting vibration would damage its mechanical seal. The ruined pump mechanical seal would then allow hazardous process fluid to escape from the pump case into the environment.

Alternately, in the event of a loss of instrument air pressure, the APO valve would shut. The pump discharge flow would also stop. The pump internal components would gradually overheat. The pump's seal faces would dry out and damage the softer carbon face. This is also bad. However, cavitation ruins seals much faster than overheating. If this control valve is designed for APO, it gives the operators more time to safely shut off the pump than if the control valve was APC.

Instrument air failure is always going to be bad news. However, we must select the path of least evil, which in this case is to have the control valve close upon loss of instrument air pressure, that is, APO. We do not need to be concerned about overpressuring the piping downstream of the centrifugal pump, as by law the piping downstream of the pump must be rated for the maximum possible pump discharge pressure (see API Boiler Code).

If we are dealing with a positive displacement pump, I would draw a different conclusion. A gear type or reciprocating pump can develop extremely high and destructive pressures once the discharge control valve is closed. For such an inline booster pump, an APC valve is preferred. I have seen a reciprocating pump lose suction pressure. It made a clanging sound, but no immediate damage was apparent.

CONTROL VALVE ON THE DISCHARGE OF A COMPRESSOR

Let us assume we have a motor-driven, fixed-speed centrifugal compressor. On the discharge of the compressor there is a control valve. Should we specify APO or APC? If the instrument air pressure is lost, the APO valve would shut. The compressor discharge pressure would rise sharply, and the flow would stop. From some very nasty personal experience at my Alkylation Unit, at the Amoco Refinery in Texas City, I know what will happen. The compressor will start to surge (see Chapter 15, "Centrifugal Compressor Surge vs. Motor Over-Amping"). Surge is a destructive phenomenon that will, in the following sequence:

- Damage the rotor's thrust bearing.
- Cause the rotating wheels to hit the stationary case elements called the labyrinth seals.
- Break off a piece of the rotor that will blast through the compressor case.

Control valves on the discharge of centrifugal compressors ought to fail in an open position (APC) upon loss of instrument air. This will lead to an unfortunate loss in suction pressure, which is also bad, but not as bad as the sudden surging that happens when blocking in the compressor discharge.

In the case of a positive displacement reciprocating compressor, I would also have a control valve on the discharge fail in an open position (APC), so as to avoid overpressure of downstream piping. This could cause a piping flange to blow out or cause the downstream pipe to rupture. A ruptured pipe looks just like someone has sliced the pipe open along its length, like a peeled banana.

I cannot think of any good reason for the Process Control Engineer to locate a control valve on the discharge of any compressor or blower. Correct design uses suction throttling and/or spillback control valves (as per Chapter 15). Regardless, control valves on the discharge of compressors are not uncommon. If used, they ought to be shown on the P&IDs as APC, so as to fail open upon the loss of instrument air pressure.

PRESSURE CONTROL OF VESSELS

Typically a backpressure control valve on a vessel should fail in an open position (APC). However, as shown in Figure 11-3, this is not always true. Here the feed valve fails closed upon the loss of instrument air pressure. The feed to V-1 stops. Let's assume that the backpressure control valve on V-1 opens on air pressure loss (APC). Then the pressure in V-2 may get quite low. Since naphtha is being condensed, the temperature and vapor pressure in V-2 could become very low, even subatmospheric. Then the pressure in V-1 would fall to

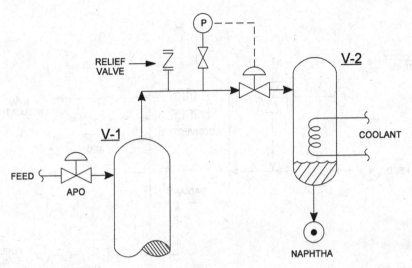

Figure 11-3 *Pressure control valve failure position depends on feed control valve failure position*

a partial vacuum. Unless V-1 was designed for such vacuum conditions, which would be unusual, V-1 could collapse. I've seen this ugly event.

Let's now assume that the backpressure control valve on V-1 closes on air pressure failure (APO). Then the pressure in V-1 is just the vapor pressure of the liquid in V-1. Of course, the possibility of overpressuring V-1 does increase if the backpressure control valve is APO. But V-1 is protected, as shown in Figure 11-3, by a relief valve from overpressure damage. However, V-1 is not protected from collapse due to excessive vacuum.

This example is a good illustration as to why the Process Control Engineer ought to be careful to think through each control valve failure position in relation to the other control valve failure positions upon the loss of instrument air pressure. This is exactly the sort of discussion that should take place during HAZOP meetings and P&ID review sessions.

FUEL GAS TO HEATERS

Figure 11-4 shows four control valves associated with a fired heater and their failure positions in the event of a loss of instrument air pressure:

- The heater outlet temperature control valve would fail in a closed position since the valve is opened with air pressure (APO). With a loss in unit controllability, the first and most important control objective is to immediately reduce heat input to the process equipment.

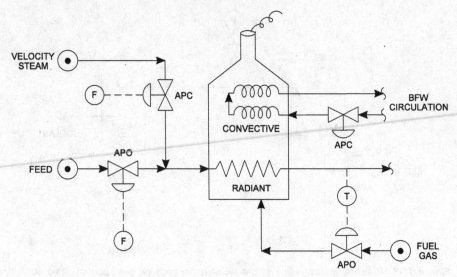

Figure 11-4 *Example of air failure valve positions for heater*

- The heater feed flow control valve would also fail in a closed position. The flow is typically coming from a centrifugal pump. As previously discussed in this chapter, the pump discharge flow would be shut in to preserve the integrity of the pump's mechanical seal.
- The velocity steam flow control valve would fail in an open position since the valve is closed by air pressure (APC). With a loss in feed flow the maximum flow of velocity steam is critical to clear residual hydrocarbons from the heater tubes. The furnace refractory lining radiates heat for many minutes. This radiated heat could promote coke formation inside the furnace tubes due to the thermal cracking of the residual hydrocarbons.
- The boiler feed water (BFW) control valve would also fail in an open position (APC). The residual radiant heat released from the refractory walls in the radiant section could damage the convective section. Continuing the BFW circulation would keep the convective section reasonably cool.

The general objective is to minimize the heat input and maximize the heat extraction from process equipment during a plant emergency such as loss of instrument air pressure. For a distillation tower, as shown in Figure 11-5, we would cause the following control valves to close (APO) during such an emergency:

- Steam flow to the reboiler to minimize heat input to the tower
- Bottoms level to avoid high pressure vapor in the tower from blowing into a storage vessel

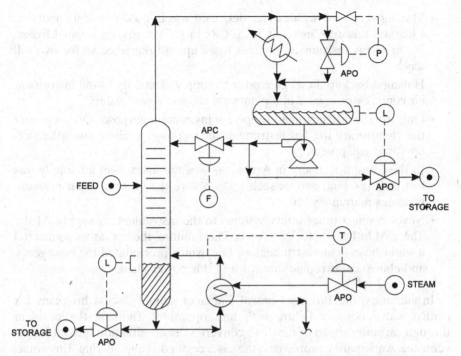

Figure 11-5 *Example of air failure valve positions for a distillation tower*

- Reflux drum level to prevent high-pressure vapor from blowing into a storage vessel
- Hot vapor bypass valve would also close (APO) to maximize the condenser heat removal.

The reflux pump flow control valve would, on the other hand, fail in an open position (APC). The reflux would be partly revaporized by the heat in the tower. Continuing the reflux as long as possible would maximize the potential to extract heat from the tower and to pass this heat on to the overhead condenser. The revaporization of flowing reflux would help prevent the reflex pump from cavitation for some reasonable period of time before the operators manually shut down the reflux pump.

AVOID LOSS OF AIR PRESSURE

Instrument air failure on a process unit is one of the most dangerous aspects of process plant control. I've always tried to avoid this by:

- Providing a backup source of nitrogen as discussed above
- Having several air compressors, each only partly loaded

- Making very sure my air dryer desiccant was in good physical condition. I learned this ugly lesson at Texas City in 1974 during an unusual freeze. Water in my instrument air lines froze up and remained so for two full days.
- Having a backup diesel generator to supply electricity to the instrument air compressor in case of a plantwide electric power failure
- Only use the instrument air supply for instrument purposes. Do not permit the "temporary use" of instrument air to power tools or run other air-operated equipment.
- In the Coastal Refinery in Aruba, 90% of the instrument air supply was lost to leaks. Find and fix such leaks before the instrument air pressure becomes marginally low.
- Never connect other utility systems to the instrument air supply. At the Three Mile Island Nuclear Power Plant failure, the operators connected a water hose to an instrument air line, which precipitated the emergency and almost catastrophic shutdown of this power plant.

In summary, the Process Control Engineer should discuss his plans for control valve position failure with unit operators. Thinking the problem through carefully ahead of time can convert a deadly situation into an inconvenience. Anticipating problems is the essence of good engineering. The worst-case scenario is that failure mode that creates a positive feedback loop. The most infamous and deadly example of this in world history is the nuclear power plant fiasco at the Soviet #3 power station in Chernobyl in 1986. This was the perfect example of Process Control Engineering at its worst. It directly led to the demise of the communist system in Russia and of the USSR's empire.

12

Override and Split-Range Process Control

Sometimes process variables must interact in a variety of complex ways such as cascade control, override control, and split-range control. I will explain the difference between these three concepts.

CASCADE CONTROL

Let us assume we are flow controlling out of a vessel. However, our primary objective is to hold the vessel level steady at 50%. The flow control valve is continuously reset to maintain the 50% level. We say that the vessel's indicated level is cascaded to the flow control valve. In the sense of process control, this sort of cascade control is quite similar to ordinary level control.

OVERRIDE CONTROL

A boiler is being fed from a deaerator as shown in Figure 12-1. The water level in the deaerator is rapidly sinking (see Chapter 20, "Steam Quality and Moisture Content"). The boiler feed water charge pump is about to lose suction and cavitate. The pump's mechanical seal will be damaged by the vibration caused by the pump's cavitation. The fuel gas to the boiler will be tripped off because of the low feed water flow. However, to protect the pump from

---

Troubleshooting Process Plant Control, by Norman P. Lieberman
Copyright © 2009 John Wiley & Sons, Inc.

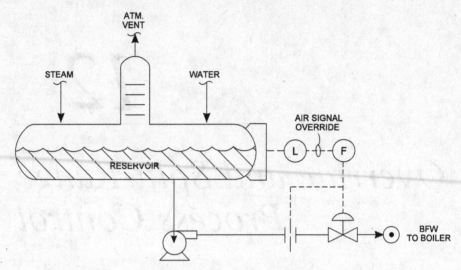

Figure 12-1 *Override pressure control on a boiler feed water deaerator*

damage, the signal from the deaerator liquid level overrides the flow signal to the control valve at the discharge of the boiler feed water charge pump. This override of the flow control will not happen until the low water level in the deaerator is activated.

SPLIT-RANGE CONTROL

Figure 12-2 shows that the pressure in the reflux drum can be maintained by either valve "A" or valve "B." Once valve "A" was completely shut, the pressure in the reflux drum would fall below its set point pressure. As this happened, valve "B" would open. Natural gas would then flow into the reflux drum to keep the drum pressure from falling below the set point pressure. The two control valves, "A" and "B," work in series to fulfill the same objective. The range of operation of each valve depends on the amount of noncondensable gas in the tower overhead and the reflux drum temperature, hence the name "split-range control." Holding a pressure with natural gas makeup represents poor process control practice as explained in Chapter 5, "Distillation Tower Pressure Control."

Split-range control employs two control valves. Both valves are controlled from a single parameter sensing point, for example, the reflux drum pressure transmitter shown in Figure 12-2.

Override control employs two parameter sensing points. Both sensing points are trying to control the same control valve, for example, the control valve on the discharge of the pump shown in Figure 12-1.

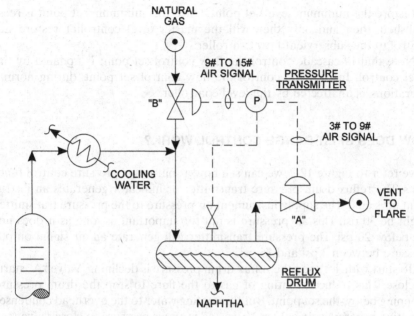

Figure 12-2 Split-range pressure control of a distillation column

Split-range control is like a good marriage. One partner may be doing 90% of the work, but both partners are occasionally going to share the work.

Override control is like a bad marriage. One partner plays a potentially dominating role, even though the other partner is doing all the work.

Cascade control is more like my marriage. I do the best I can, but my wife Liz constantly and lovingly recalibrates my efforts. She dampens down the extremes in my behavior so as to promote a stable relationship and home life.

Cascade control differs from override control in several ways. For example, a flow control is being reset by a level. If the flow control is drawing the level down below the 50% level set point, the level control will signal the flow control to cut back its signal. If the flow control is permitting the level to rise above the 50% level set point, the level control will signal the flow control to increase its signal. This goes on all the time, unless the panel board operator switches the cascade control from automatic to manual, which means the control is now just on direct flow control.

Override control is different from cascade control. For example, a flow control is being overridden by a low level. The flow control changes to hold some flow set point without any interference from the low level override. During normal operations the level override feature is never used and should never be used. However, should the level fall below some low level set point, then the level override will seize control of the process from the flow controller to restore the minimum level set point. Once this minimum set point is reestablished, then and only then will the master level controller restore full control to the subservient flow controller.

Note that in cascade control the flow control set point is updated by the level control. In override control the flow control set point, during normal operations, is unaffected by the level controller.

HOW DOES SPLIT-RANGE CONTROL WORK?

If we refer to Figure 12-2, we can see how a split-range pressure control functions. The reflux drum pressure transmitter or indicator generates an instrument air signal. The supply instrument air pressure to the pressure transmitter might be 30 psi. This air pressure is not too important as long as it does not get below 20 psi. The pressure transmitter will generate an air signal output pressure between 3 psi and 15 psi.

To start with, let's say the reflux drum pressure is declining. Valve "A" starts to close. This reduces venting of gas to the flare to stop the drum pressure dropping below the set point. But the cooling water to the overhead condenser is getting progressively colder. Valve "A" is an air pressure to close valve (see Chapter 11, "Control Valve Position on Instrument Air Failure"). Normally such a control valve will be 100% open with 3 psi of instrument air pressure and shut with 15 psi of instrument air pressure. However, this control valve has been calibrated to be 100% open with 3 psi of instrument air pressure and shut with 9 psi of instrument air pressure. Because of the colder cooling water, the off-gas production from the reflux drum drops to zero.

Valve "A" with 9 psi of instrument air pressure is closed. The falling reflux drum pressure continues to generate a larger instrument air pressure signal from the drum pressure transmitter above the 9 psi that has shut valve "A." Valve "B" now starts to open. Valve "B," unlike valve "A," is an air pressure to open type valve (as described in Chapter 11). As the reflux drum pressure continues to drop because of the colder cooling water, valve "B" continues to open. An air pressure to open valve is normally shut when the instrument air pressure is 3 psi and fully open when the instrument air pressure is 15 psi. However, this control valve has been calibrated to be closed with 9 psi of instrument air pressure and 100% open with 15 psi of instrument air pressure.

Natural gas now flows into the reflux drum and dissolves in the naphtha product. This increases the vapor pressure of the naphtha and thus restores the reflux drum pressure to its original set point. The calibration of valves "A" and "B" should be such that both valves are completely closed at 9 psi instrument air output pressure generated by the reflux drum pressure transmitter. I suppose that this is theoretically possible. In practice, it doesn't seem to work. That is, natural gas will leak to the flare vent even with an 8 psi air signal from the reflux drum pressure transmitter.

I have selected this particular example of split-range control because it's easy to explain. In practice, this is an especially wasteful and inefficient form of distillation tower pressure control, as I discuss in Chapter 5.

Mechanically there is no real difference between valve "A" and valve "B." To change a control valve from air pressure to open to air pressure to close is simple:

- An air pressure to close valve, like valve "A" in Figure 12-2, will have the instrument air signal tubing connected on top of the valve diaphragm.
- An air pressure to open valve, like valve "B," will have the instrument air signal tubing connected beneath the valve diaphragm.

A ¼-inch diameter length of copper tubing connected above or below the diaphragm indicates the type of control valve.

SAFETY TIPS

An over-speed trip on a steam turbine is an example of override control. Should the turbine speed exceed 3750 rpm, the motive steam will be shut off. Normally, the governor steam control valve is set to hold the turbine speed at 3600 rpm. However, if the governor valve sticks open because of salt deposits, then the trip valve overrides the governor and blocks in the motive steam flow (see Chapter 17, "Steam Turbine Control").

If the flow to a fired heater gets too low, a low-flow trip will shut off furnace fuel. Normally the fuel gas rate is controlled by the heater outlet temperature (TRC). But should the heater charge rate get too low, the fuel gas trip overrides the TRC and fuel gas flow to the burners stops completely (see Chapter 22, "Alarm and Trip Design for Safe Plant Operations" for examples).

Normally, lube oil to my alkylation unit refrigeration centrifugal compressor in Texas City was supplied by a turbine-driven lube oil circulating pump. However, should the lube oil pressure get too low, the backup electric motor driven lube oil pump would trip on automatically to sustain the minimum lube oil minimum pressure set point (see Chapter 19).

Override control can be multivariable. Too high a fuel gas rate to a furnace may be overridden by an excessively hot firebox temperature transmitter. Too low a fuel gas may be overridden by a low furnace feed flow transmitter. Cascade control may also be multivariable. For example, a pressure control can cascade down to a level control, which then cascades down to a flow control valve on the discharge of a pump.

In summary, complex control schemes are fine, but they have to work in practice, not only on paper. The distillation tower pressure control scheme depicted in Figure 12-2 is an example of a widespread and common control method that is in practice quite objectionable.

ENHANCED CONTROL VALVE SAFETY

With the development of electronics and distributed control, it is becoming increasingly common to indicate in the control room the actual control valve position. I'll define several terms:

- **I/P**—This device converts an electronic signal from the panel into a pneumatic (air) signal to a control valve. The control valve then functions using air pressure in the manner I've described above.
- **Transducer**—Converts an air pressure signal to an electronic signal.
- **Positioner**—This device transforms the electronic signal from the control room to an air signal. Also, it measures the actual valve position (that is, the position of the control valve's stem) and transmits this position back to the panel in an electronic form.

The electronic positioner works by using the transducer and I/P devices. In the field it looks like a large cylinder or can set on top of the control valve. On top of this cylinder you can see an indication of the actual valve position. This is not the air signal to the control valve diaphragm, but the real control valve stem position.

The output from the electronic positioner may then be used to correct or verify the control valve position that is indicated in the control room, and which has been generated to apply pneumatic air pressure to the control valve diaphragm. In effect, the control valve in the field is reporting back to the panel as follows:

- "Yes, I have checked the valve stem position and everything is correct."

Or, if the control valve is not working correctly:

- "No, the control valve is failing to obey the signal from the panel."

Another method to upgrade the reliability of a control valve is to have two or three different measurements of the same process variable generating multiple outputs to control a single control valve. The idea is to move the control valve to a safe position, even if one of the three measurements is requiring the valve to move to an unsafe position. For example, a level control valve on a pump discharge may be trying to open the control valve. But the pump's suction pressure transmitter observes that the pump is about to cavitate because the pump's suction screen is plugged. The transmitter then stops the discharge control valve from opening, even though the indicated liquid level is rising.

This is another type of override control that I describe in detail in Chapter 12, "Override and Split-Range Process Control." The difference is that three or more variables can be taken into consideration, and that electronic signals are used.

13

Vacuum System Pressure Control

Lubrication oils, medicinal and mineral oil, baby oil, paraffin wax for candles, and microcrystalline wax for furniture polish are all produced by vacuum distillation of petroleum residues. The base stocks for many (if not most) cosmetics is largely just hydrotreated heavy-vacuum gas oil produced in petroleum refineries. Especially for automotive lubricating oils and paving asphalt used to make roads, efficient vacuum distillation is critical. A few mmHg fluctuation in the operating pressure will severely diminish fractionation efficiency in any vacuum tower.

The control of vacuum tower pressure is complicated by the extremely nonlinear performance of the converging-diverging steam jets. Some vacuum systems use a simple steam jet consisting of just a diverging section, often called a "hogging jet"; only relatively poor vacuums can be produced. These jets are inefficient and are normally not intended for continuous operation.

Two vacuum liquid ring seal pumps in series will produce an excellent vacuum. These pumps are really positive displacement compressors. They are energy efficient, and it is easy to control their suction pressure. Highly reliable, they only have one serious drawback. They are very expensive to purchase and install. Each liquid seal ring pump is like a miniature process plant with heat exchangers, pumps, and control valves.

For most industrial applications, the converging-diverging steam jet ejector is usually employed with two or three units in series. It is the standard way to develop a deep vacuum. The nonlinear performance of such steam jets or

Troubleshooting Process Plant Control, by Norman P. Lieberman
Copyright © 2009 John Wiley & Sons, Inc.

ejectors greatly complicates vacuum column pressure control. The problem lies with the converging portion of the jet. I will explain this problem in some detail.

When high-velocity steam exhausts from the steam nozzle shown in Figure 13-1, the steam pressure of 100 psig is entirely converted to kinetic energy, so that a vacuum of slightly less than 10 mmHg is produced, as shown in the figure.

For clarity, I will be using the absolute scale of vacuum measurement:

- Atmospheric pressure at sea level = 760 mmHg
- Full vacuum = 0.0 mmHg
- 29.97 inches Hg vacuum = 0.0 mmHg
- 0.1 bar (absolute) = 76 mmHg

A typical design ratio for moles of motive steam per mole of vapor to be compressed is 4 to 1. The combined effluent from the mixing chamber enters the converging portion of the diffuser, where it accelerates to a velocity above the speed of sound. As the vapors pass through sonic velocity, a pressure wave front is created, which I call the sonic boost. This pressure wave front compresses the vapor from 10 mmHg to 40 mmHg for a compression ratio of 4 to 1.

After the vapors at 40 mmHg leave the diffuser throat shown in Figure 13-1, they deaccelerate in the diverging portion of the jet. The reduced kinetic energy or velocity is partly converted to compressing the vapors from 40 mmHg to 100 mmHg, for a compression ratio of 2.5 to 1. I call this compression ratio the velocity boost. The overall compression ratio of the ejector is 100 mmHg divided by 10 mmHg or 10 to 1. About two-thirds of the overall compression ratio is due to the sonic boost.

Let's say I have placed a back pressure control valve of the discharge of the jet. This is not an acceptable method of control, so this is only a hypotheti-

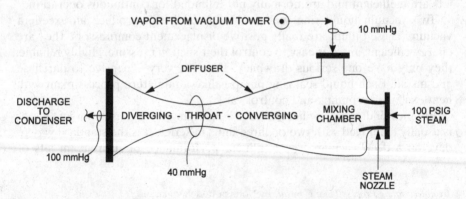

Figure 13-1   Converging-diverging vacuum steam ejector

cal discussion. As I close this control valve, the discharge pressure of the jet may increase from 100 mmHg to 110 mmHg. The volume of vapor passing through the jet discharge is reduced by 10% (volume varies inversely with pressure), and the velocity also slows by 10%. The velocity boost will be reduced linearly by roughly 10%. The pressure of the vapor in the diffuser throat will now increase because of two factors:

- The increase in the jet discharge pressure to 110 mmHg
- The reduction in the velocity boost by about 5 or 6 mmHg

The pressure in the diffuser throat has risen to about 55 mmHg. However, as long as the velocity in the diffuser throat slightly exceeds the speed of sound, the jet is still said to be in its "critical mode of operation." I would rather say the jet is still developing its full sonic boost of 4 to 1. As the sonic boost accounted for the major portion of the overall compression ratio, the affect on the vacuum tower top pressure is really small, often too small for me to measure.

However, if you persist in closing the discharge control valve past a point, you are in for a shocking and unpleasant surprise, for at a certain jet discharge pressure the vapor velocity in the diffuser throat will fall slightly below the speed of sound or sonic velocity. Then the sonic boost vanishes completely and utterly in 1 second. The suction or jet inlet pressure might suddenly increase from 12 mmHg to 60 mmHg. The discharge pressure at which this surprising loss in vacuum jet performance occurs is called the "critical jet discharge pressure." It's a design characteristic of any converging-diverging steam ejector.

Operators say that this sudden change in the jet's performance results in the vacuum "breaking." They ought to say the jet has lost its sonic boost and has been forced out of its critical mode of operation. This happens because the jet's discharge pressure exceeded its critical design discharge pressure.

Our objective as Process Control Engineers is to select a mode of control that responds proportionately to a change in a parameter's input. Locating a control valve on the jet's discharge flow does not answer to this requirement, because of the sudden loss in sonic boost.

FACTORS AFFECTING LOSS OF SONIC BOOST

There are other factors that can cause a nonlinear response in jet performance due to the sudden and total loss of the sonic boost. I've experienced them all:

- Overloading due to a large air leak
- Overloading due to excessive cracked gas flow
- High discharge pressure due to overloading of the downstream condensers

- Low motive steam pressure to the jet
- Excessive moisture in the motive steam
- Erosion of the jet's steam nozzle (see Fig. 13-1)
- High motive steam pressure to the jet
- Plugged barometric drain lines from the discharge condenser to the seal drum
- Internal damage to the air baffle in the jet's discharge surface condenser

Any control method that causes a transition from the normal jet performance with the sonic boost to handicapped jet performance without the sonic boost will result in a highly nonlinear response in the vacuum tower pressure. This in turn will make efficient fractionation quite improbable. For conventional vacuum ejector systems there is no practical solution to this problem.

SPECIALTY VACUUM EJECTORS

Several years ago I was working on a vacuum tower waste gas recovery system for Chevron at their Pascagoula, Mississippi refinery. They had purchased a vacuum jet system in which a control valve placed on the motive steam supply could operate over a very wide range and still produce a proportionate change in the vacuum tower top pressure. Thus for a new vacuum system installation I will alert the Process Control Engineer that controlling motive steam pressure can be used as an effective means of vacuum tower top pressure control. However, I will also note the following:

- These ejectors had been specially purchased for this purpose and are not likely to be the typical jets you will encounter in your plant.
- How it all worked out in the field I do not know. Apparently, this was also the first time that the Chevron Corporation had used such equipment, as they pretested it first under laboratory conditions.

As I've stated in the introduction to this text, I will only recommend control applications and process equipment based on first-hand experience. This precludes this particular jet application.

THROTTLING ON MOTIVE STEAM

I tried to manually control the motive steam pressure to the jet at the Coastal Refinery in Aruba. As I reduced the motive steam pressure from 160 psig to 130 psig, the vacuum became very slightly better. Likely I was unloading the downstream condenser with less motive steam to condense. At 120 psig the

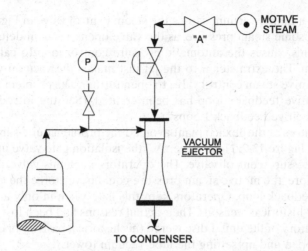

Figure 13-2   *Throttling on motive steam supply sometimes works well in vacuum tower pressure control*

vacuum suddenly broke and the vacuum tower pressure jumped up. I had lost the sonic boost. Or as the jet vendors say, "I had been forced out of the critical mode of operation by low motive steam pressure."

Of somewhat greater success was the longstanding control method used by the Texaco (now Motiva) Refinery on their giant lube oil vacuum tower in Port Arthur, Texas. The jet shown in Figure 13-2 was designed to develop a sonic boost. But the jet had not worked with its sonic boost for a very long time, if ever. As long as an ejector is only developing its velocity boost, it is said to be in its "throttling mode of operation." This means that a small closure of the motive steam inlet control valve will result in a small reduction (i.e., increase in pressure) in the tower vacuum. Increasing the motive steam pressure did improve vacuum, but only to a point.

POSITIVE FEEDBACK LOOP

Even when operating in the throttling mode (without the sonic boost) too much steam pressure can sometimes reduce the jet's compression ratio. This may happen because:

- The downstream condenser is overloaded.
- The steam nozzle in Figure 13-1 is worn out because of wet steam or lack of maintenence.
- The motive steam pressure is above its design pressure.

Normally a falling vacuum causes the steam control valve in Figure 13-2 to open. But at some steam pressure, as the valve opens, vacuum deteriorates in the tower. This causes the automatic pressure control loop to call for more motive steam. The extra steam to the jet just makes the vacuum worse. This forces the motive steam control valve to open further. The problem feeds upon itself. A positive feedback loop has been created. (See the Introduction, "A History of Positive Feedback Loops")

The operators at the Texaco plant avoided the problem quite simply. Again, let's refer to Figure 13-2. Note valve "A," the isolation gate valve upstream of the steam pressure control valve. The operators kept this valve about 25% open. Therefore, the motive steam pressure could never force the ejector into a positive feedback loop. Operators kept this gate valve in only a 25% open position for "historical" reasons. Their actual reasons had been lost in the dim operating history of the unit. I discovered this historical basis by opening valve "A" to half open and up-setting the lube vacuum tower.

SPILLBACK PRESSURE CONTROL

I rather favor throttling on the motive steam supply to a jet, not because process control is better than the alternate method of spillback control but because steam is saved. For example, reducing the motive steam pressure to a jet from 120 psig to 60 psig saves about half the steam. Regardless of my opinion, spillback control, at least in the petroleum refinery industry, is the most common sort of vacuum tower pressure control. And as long as the jet is not overloaded by excessive gas flow and forced out of its sonic boost, this mode of control works extremely well. The jet suction pressure is entirely predictable. It is defined by the vendor's performance curve for the jet, which plots the vapor load to the ejector against its intake suction pressure.

In spillback control, gas from the last stage jet is spilled back through the vacuum tower overhead vapor line to the first stage inlet, as shown in Figure 13-3. If the jet is not developing its sonic boost, spillback control also works in a linearly responsive manner. It's just the transition from the critical mode to the throttling mode that cannot be tolerated.

In 1989 I visited the Chevron plant in Richmond, California. I was trying to sell a Process Control Study to promote Energy Conservation for $50,000. Chevron was not interested. They said they already knew everything. So I offered them a free sample. "Show me any part of your plant and I will instantly save energy by improving the process control."

They drove me out to their super-gigantic lube oil vacuum tower. It had three first-stage jets running in parallel. The spillback (Fig. 13-3) was so far open that the panel board operator could not properly control the vacuum tower pressure. The control valve was not in the linear portion of its operating range, and fractionation efficiency was impaired (see Chapter 10, "Sizing Control Valves").

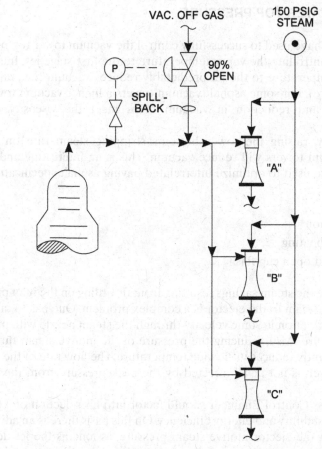

Figure 13-3 *Spill-back pressure control. Too many jets in service*

Each of the giant first-stage jets was using 18,000 lbs/h of 150 psig steam. The outside operator and I took jet "C" out of service. The spillback valve closed to 50%, and the vacuum tower pressure control became stable.

"Well," I said, "I just saved 22 mm BTU/h worth of 150 psig steam by using my *Enhanced Process Control Program*. Maybe now you'll be more interested in purchasing my $50,000 Process Control Study."

"Bullshit, Lieberman, anyone could have done that," the Chevron engineers responded.

Yes, but that's my point. Anyone could have done that. Anyone who understood the interaction between the process equipment and the process controls.

VARYING TOWER TOP PRESSURE

Other ways I have used to successfully control the vacuum tower top pressure depend on controlling the vapor temperature to the first-stage jet. Increasing the vapor temperature to the jet predictably reduces vacuum (i.e., raises the tower pressure). For some asphalt vacuum towers a higher vacuum tower top temperature and reduction in vacuum helps meet the viscosity asphalt specification.

Alternately, raising the tower bottoms stripping steam rate on paving asphalt vacuum towers will reduce vacuum. This is an interesting and useful method I have used to optimize interrelated paving asphalt specifications:

• Viscosity
• Penetration
• Loss on heating
• Cleveland open cup flash

Calculating the steam savings resulting from throttling on the inlet pressure of the motive steam to the ejector is a complex problem. One has to take into account the variation in sonic velocity through the steam nozzle with pressure and temperature. Also, reducing the pressure of the motive steam through a valve significantly reduces its flowing temperature. The flow rate of the motive steam, however, is not at all affected by the back pressure from the mixing chamber.

The Process Control Engineer should factor into his selection of a control scheme both stability and energy efficiency. On this basis there is an advantage for control of the ejector motive steam pressure, as long as the jet does not suddenly lose or restore the sonic boost. Or, as explained in the vendor literature, as long as the jet does not leave or enter its "critical mode of operation."

14

Reciprocating Compressors

Reciprocating compressors are simple positive displacement machines. The gas is pushed out of a cylinder into the discharge line by the force of the piston head. As the piston reverses its direction of travel, new gas is drawn into the cylinder from the inlet line.

Some reciprocating compressors are driven by electric motors. These machines run at a fixed speed. This constant speed characteristic creates a control problem when the maximum gas flow is not required. Other reciprocating compressors are driven by gas engines. The gas engines are quite similar to an automobile engine, except that the fuel is natural gas rather than gasoline. Reducing the capacity of a gas engine-driven reciprocating compressor may be as simple as slowing down the engine.

HIGH DISCHARGE TEMPERATURE TRIP

Much of my experience in dealing with the control of reciprocating compressors was gained in Laredo, Texas. George Garza and I would spend all day driving around the natural gas fields south of Laredo adjusting reciprocating wellhead compressors. Often, especially during hot weather, we would find a wellhead compressor had tripped off because the discharge temperature had exceeded the trip point. This puzzled me. The trip temperatures were typically in the range of 350°F. All components of the compressor exposed to the

Troubleshooting Process Plant Control, by Norman P. Lieberman
Copyright © 2009 John Wiley & Sons, Inc.

discharge flow were made of carbon steel, which does not lose any structural strength until a temperature in excess of 750 °F is reached. Why, then, the 350 °F limit?

My research on this question revealed that the high temperature trip on the compressor discharge gas flow was not intended to protect the compressor from a high temperature. It was to protect the compressor from piston rod failure.

The logic was:

1. Either a falling suction pressure or a rising discharge pressure increases the pressure rise across the compressor. Also, the compression ratio increases.
2. The pressure difference between the inlet and outlet of the compressor creates a pressure differential across the piston head (see Fig. 14-1).
3. The differential pressure across the piston head (300 psig – 60 psig = 240 psi) multiplied by the area of the piston head (in square inches) is the force (in pounds) that is exerted on the piston rod.
4. The temperature rise is proportional to the compression ratio. Thus the high discharge temperature is an indirect measurement of excessive piston rod loading.

During hot days in Laredo, the inlet gas to the compressor might increase by 20–30 °F above normal. The discharge temperature would increase by the

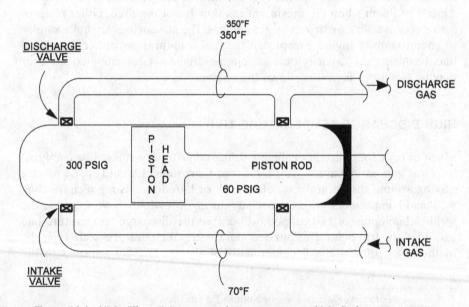

Figure 14-1 High differential pressure trips compressor on high discharge temperature

same amount, even though the compression ratio was constant. The piston rod was not affected by the hotter gas. But still the hotter gas would cause the compressor to trip, to protect the compressor's piston rod from failure due to excessive loading.

This all makes zero sense. The high temperature trip was just a cheap, easy way for the vendor to protect his equipment against catastrophic piston rod breakage. The wellhead compressor would shut down on the hottest days of the year, just when natural gas demand was greatest from utility companies trying to generate electricity for air conditioners. Just when gas sales on the spot market were commanding the best price.

There is a general lesson to be learned from this story. So often, artificial control limitations are imposed on a facility. These limitations, having never been challenged, become part of the operating culture of a plant. The reasons for the control limitation may have long been forgotten, but the limitation itself lives on. The Process Control Engineer should identify, and where appropriate discard, control parameter limitations that have no technical, but only historical, basis.

MEASURING PROCESS FLOWS

In 1993, I designed a new hydrogen plant feed compressor for the coastal refinery in Aruba. The aftermath of this project provides a general example of the importance of understanding how a process flow variable is measured. In this case the variable was a gas flow.

Flow is measured according to the following equation:

$$\text{Delta } P = K \cdot (DEN)(V)^2 \qquad (Equation\ 1)$$

where:

- Delta P = pressure drop through orifice plate
- K = orifice coefficient
- DEN = density of the fluid
- V = velocity through orifice plate

Solving for velocity in the above:

$$V \text{ is proportional to } \frac{(\text{delta } P)^{1/2}}{(DEN)^{1/2}} \qquad (Equation\ 2)$$

The velocity through the orifice plates multiplied by the orifice plate area is the volume of gas flowing in the pipeline. Note from *Eq. 1* that if the density of the gas drops by a factor of four, the delta P of the gas flowing through the orifice plate will also drop by a factor of four. Alternately, if the velocity or volume of gas flowing through the orifice plate drops by a factor of two, the delta P of gas flowing through the orifice plate will drop by a factor of four. It's rather easy to misinterpret a reduction in gas density as a reduction in gas flow; unless you understand how flows are measured, which the engineers on the island of Aruba did not.

I had been given two different feed composition design cases for the feed gas compressor:

- **Case 1**—Purge hydrogen gas, with a molecular weight of 8 lbs/mol (8 MW)
- **Case 2**—Plant fuel gas, with a molecular weight of 32 lbs/mol (32 MW)

It is the fundamental nature of reciprocating compressors that molecular weight has only a tiny effect on the volumetric capacity of the compressor, meaning that the compressor will pump the same volume of gas whether the gas has an 8 molecular weight or a 32 molecular weight. (Caution: This logic does not apply to centrifugal compressors.)

I arbitrarily selected case 2, the 32 MW fuel gas design case. When the new compressor was commissioned with fuel gas, it delivered the design flow of $400,000 \, \text{ft}^3$ per hour. When the operators switched to the 8 MW purge hydrogen, the panel flow indicator dropped in an alarming fashion, to $200,000 \, \text{ft}^3$ per hour. The smaller feed gas flow could not sustain the operation of the hydrogen plant for very long. Thus, seeing a rapid loss in the feed rate, the panel operator switched the feed gas back to fuel gas.

A large engineering meeting was held with 20 attendees to discuss my apparent misdesign of the new reciprocating compressor. At the meeting a letter was produced from a reciprocating compressor valve vendor. The letter stated that new valves were required for the compressor to handle the 8 MW hydrogen-rich gas. Also, the letter suggested that my design based on the 32 MW gas was too restrictive and inconsistent with the 8 MW feed gas operation.

When many people first meet me, they think they will not like me. But after people really get to know me, they become quite sure that they don't like me. For example, I did not rationally and calmly explain the problem of the varying molecular weight and its apparent effect on the gas flow. I just picked up my slide rule and flew home to New Orleans.

The 32 new compressor valves were installed at an all-up cost of $500,000. When the revised machine was restarted, it behaved just as before. Apparent gas flow dropped by half, when the feed gas was switched from 32 MW to 8 MW. Standing at the back of the compression skid watching the futile activity, I silently observed this human comedy.

The Aruba engineers had neglected to correct the gas flow, as measured by the orifice meter, for density. Had they used the equations shown above, they would have realized that the actual flow of gas—purge hydrogen or fuel gas—had never changed. It had always been 400,000 ft³ per hour. The gas flow had never actually dropped to 200,000 ft³ per hour. It's just that the gas density had declined by a factor of four. The same thing happens with liquids. For example, we are metering 100 GPM (gallons per minute) of water flow. The density of water is 1.0 SG. Now we switch to metering gasoline at a flow of 100 GPM, with a 0.80 SG. The apparent metered flow will be 90 GPM:

$$100 \, \text{GPM} \div \frac{(1.00 \, \text{SG})^{1/2}}{(0.80 \, \text{SG})^{1/2}} = 90 \, \text{GPM}$$

Why did the reciprocating compressor valve vendor offer to sell the $500,000 (U.S.) new compressor valves for the purge hydrogen? I suppose that not everyone is as honest and trusting as the happy island people in Aruba.

FLOW CONTROL

If a reciprocating compressor is driven by turbine or gas engine, flow control is achieved by simply reducing the compressor speed. To reduce the flow through a fixed-speed reciprocating compressor is far more difficult. There are several control techniques available. I'll list them from worst to best:

- Discharge throttling (worst)
- Suction throttling
- Spillback
- Valve unloaders
- Adjustable head end unloader (best)

Throttling—Discharge or Suction

I've never observed discharge throttling on a compressor. For a reciprocating compressor discharge throttling would only have a small effect on gas flow rate. It would not work too well. Suction throttling is widely used, effective, and the correct way to control the flow and upstream pressure for a centrifugal compressor, as discussed in Chapter 15, "Centrifugal Compressor Surge vs. Motor Over-Amping." Suction throttling for a reciprocating compressor is also effective in reducing flow, but it wastes energy by increasing the compression ratio (discharge pressure divided by suction pressure). More importantly, suction throttling increases piston rod loading, which could cause the piston rod to break. I have never seen a design incorporating suction throttling for a reciprocating compression, but I have seen operators resort to this undesirable practice in gas field operations.

Spillback

Most reciprocating compressors' flows are controlled by recirculation through a spillback cooler. That is, the discharge is recirculated to the suction of the compressor, as shown in Figure 14-2. The cooler is needed to remove the heat of compression from the recirculated gas. The heat that is removed represents wasted compression work or energy. It is this energy waste that is undesirable. Otherwise, spillback control works fine; thus its widespread use.

Valve Unloaders

I object to the use of spillback control on the principle of energy waste. In theory valve unloaders do not waste energy. In reality they do. Valve unloaders are actually valve disablers. Steel fingers reach into the valve, through the valve cap cover, and prevent the valve plate from closing. I realize that unless you have seen a reciprocating compressor valve you will not understand from the mechanical perspective what I have written. But from the Process Control Engineers perspective, only two things need to be understood:

1. When a valve disabler or unloader is activated on a compression cylinder, that portion of the cylinder effected will no longer pump any gas.
2. However, the compression cylinder still consumes about 20% of the energy it consumed when it was in service.

Evidence of this energy waste is that the idled compression cylinder gets quite hot.

Valve unloaders are also bad in that when a section (i.e., 50%) of a cylinder is idled, the gas flow drops by that capacity of the section of the compressor cylinder. If there are two cylinders working in parallel, the operator reduces the gas flow step-wise, in 25% increments. This large a change is excessive for the control of many process operations.

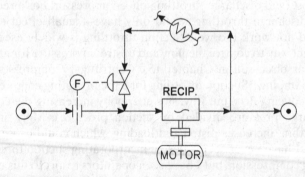

Figure 14-2   Flow control with spillback is an energy-wasting design

Adjustable Head End Unloaders

This is the correct way to control gas flow through a reciprocating compressor. Refer to Figure 14-3. The piston head is shown at the end of its travel. That is as close as it gets to the cylinder head. The amount of gas trapped between the cylinder head and the piston head is labeled "Vol" on the figure. The bigger "Vol," the smaller the capacity of the cylinder. "Vol" can be increased by use of an adjustable head end unloader device (not shown). Again, I will not explain mechanically how this works. But from the control engineer's perspective:

1. The adjustable head end unloader can reduce the capacity of a cylinder up to about 30%, without wasting energy.
2. Capacity reduction occurs in small increments, resulting in good flow control.

All reciprocating compressors can be retrofitted with these adjustable pockets. The pocket looks like a large valve handle that bolts onto the cylinder

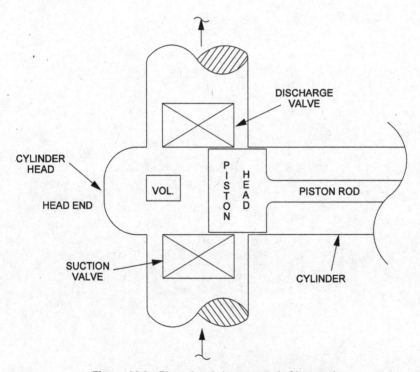

Figure 14-3 Piston head shown at end of its travel

head. The disadvantage of the adjustable pockets are that they are very expensive. Also, they can only partly reduce the capacity of the cylinder head end compression, and they do not effect compressor capacity for the crank end. Hence, a reduction in flow only up to 25–30% can be achieved with the adjustable head end unloading pocket.

15

Centrifugal Compressor Surge vs. Motor Over-Amping

Let's begin the discussion of centrifugal compressors with a story about a sulfur recovery plant in Venezuela. I was working in Maracaibo for a PDVSA refinery. The project pertained to their sulfur recovery plant air blower. Air is supplied at a precise pressure and carefully controlled flow to oxidize hydrogen sulfide into water plus elemental sulfur:

$$H_2S + O_2 = H_2O + S_{(8)}$$

The air blower is a large, low-head constant-speed centrifugal compressor, driven by a 440-volt, three-phase electric motor operating at 3600 rpm. The majority of motors in process plants are of this type. As shown in Figure 15-1, the blower is venting excess air to the atmosphere to maintain a constant pressure at the orifice plate measuring combustion air flow. If the pressure of the air flow to the orifice plate varied, control of air flow would have to be corrected for air density. This is possible, but also complex. So, to simplify control, the pressure at the orifice plate is kept constant.

The question at the Venezuelan sulfur plant was how much energy would be saved by cleaning the air blower suction filter. My calculations contradicted the expert from Conservation Survey Incorporated, a well-known company in the field of energy conservation. The survey corporation expert had predicted that changing the filter elements at the blower suction would decrease the filter pressure drop from 20 to 4 inches of water. This, he stated, would reduce the

Troubleshooting Process Plant Control, by Norman P. Lieberman
Copyright © 2009 John Wiley & Sons, Inc.

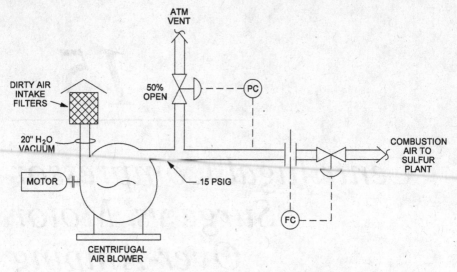

Figure 15-1 *Filter plugging reduces motor amps on sulfur plant air blower*

load on the air blower motor by 4%, that is, in proportion to the increase in the air blower suction pressure. He noted that 1 atmosphere is equal to 400 inches of water: 16 inches ÷ 400 inches = 4%.

I maintained just the opposite. I had calculated that decreasing the pressure drop across the filter would increase motor power consumption by 12%. The PDVSA technical director disagreed. He stated that restricting the air flow to the suction of the blower would increase the energy required to compress the air to the required discharge pressure.

I never like to argue. My calculations did not convince either my Venezuelan client or the survey corporation's energy expert. They both maintained that my ideas were counterintuitive. Obviously, they both said, reducing the blower's compression ratio must reduce compression work. This was just common sense. So I slid a piece of plywood across the filter's lower intake port. We then observed the following:

- The inlet pressure to the blower dropped from 20 inches of water vacuum to 24 inches of water vacuum. That is, my plywood reduced the blower suction pressure by 1%.
- The atmospheric vent valve shown in Figure 15-1 closed from 50% to 40%.
- The amperage load on the blower's motor driver dropped by 5%.

The survey corporation expert was disturbed and puzzled. My client, the PDVSA technical director, was quite pleased. He grabbed a second section of plywood and placed it in front of the blower upper intake port. The blower suction pressure slipped down to 28 inches of water vacuum. The blower suddenly began to surge in a most frightful manner! What had happened?

UNDERSTANDING BLOWER CONTROLS

When an air blower surges, air flow swings from normal to zero. This damages the blower's thrust bearing and will with time destroy the rotor. Also, the control of combustion air flow to the sulfur recovery plant becomes impossible. Figure 15-2 is the operating curve for the air blower. Note the following:

- The horizontal axis is the actual cubic feet of air flow. As the blower suction pressure drops, the number of pounds of oxygen in each cubic foot of air is reduced. Lower pressure reduces air density.
- The vertical axis is the feet of head developed by the blower. As the blower suction pressure drops, the feet of head required to reach the 15 psig discharge set point pressure will increase. The compressor, following its operating curve, will backup the curve toward the surge point. This further reduces CFM, the air flow in cubic feet per minute.
- Air flow at the discharge of the blower drops. This forces the atmospheric vent valve to close to maintain the set point pressure at the flow control orifice. Thus, reducing the blower suction pressure reduces venting of air from the blower discharge. Hence, less air is compressed, and the work required by the motor is reduced.

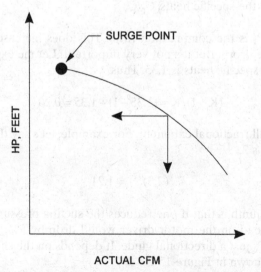

Figure 15-2 *Fixed-speed centrifugal compressor operating curve. Polytropic head vs. cubic feet per minute*

I explained to my Venezuelan colleague he had gone too far in reducing the blower suction pressure. Reducing the blower suction pressure to 28 inches of water vacuum had pushed the blower to an air flow below its surge point. Therefore, the blower had begun to surge.

The survey corporation expert was still very much puzzled. Yes, he agreed that the air flow had gone down. But the overall compression ratio had increased. Which was more important, he asked, the decreased flow or the compression ratio going up? How, he asked, would one calculate the relative effects?

THE SECOND LAW OF THERMODYNAMICS

According to the second law of thermodynamics, compression work (which is proportional to motor amperage) is calculated according to *Eq. 1*:

$$\frac{R \cdot (N \cdot T_1 \cdot K)}{(K-1)} \times \left[(P_2 \div P_1)^{(K-1)K} - 1 \right] \qquad (Equation\ 1)$$

where

R = Natural Gas Constant
N = number of moles
T_1 = suction temperature, °R
P_2 = discharge pressure, psia
P_1 = suction pressure, psia
K = ratio of the specific heats, C_p/C_v

P_2 divided by P_1 is the compression ratio. This does increase as the blower suction pressure drops. But it's not very important. Let me explain. K, for air, the ratio of the specific heats is 1.35. Thus:

$$(K-1)/K = (1.35-1) \div 1.35 = 0.26 \qquad (Equation\ 2)$$

This is a small fractional exponent. For example, let's say that the blower's compression was 2.

$$(2.0)^{0.26} = 1.21 \qquad (Equation\ 3)$$

My rule of thumb is that if one reduces the suction pressure by 10%, then the amperage load on the motor driver would drop by 5%. Don't take this rule to heart. It's just a directional guide. It depends on the shape of the head vs. flow curve, shown in Figure 15-2.

The PDVSA technical director was beginning to understand. He noted several limitations to suction throttling:

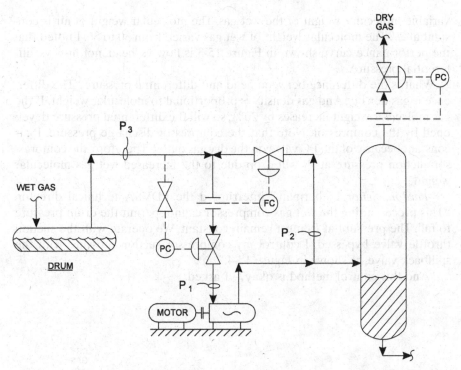

Figure 15-3 *Suction throttling pressure control for a motor-driven centrifugal compressor*

- This concept only applied to fixed-speed machines, not variable-speed turbine drivers.
- One needed to stay a safe amount above that low air flow which could cause surge.
- The atmospheric vent pressure control valve must not be completely closed.

"*Señor* Lieberman, this is *muy bueno*," said the technical director. "Now I see that the atmospheric vent is like a spillback. Just like we have on our wet gas compressor. It's also a centrifugal compressor driven by a constant-speed motor. It too has a suction throttle valve (see Fig. 15-3) like your plywood. It has a spillback valve, which is the same as the atmospheric vent on the sulfur plant blower. But *Señor* Lieberman, we always operate with the suction throttle valve 100% open. Now, maybe, I think this is *muy malo*."

EFFECT OF WET GAS MOLECULAR WEIGHT

I explained to the director that the problem with his wet gas compressor was more complex than the air blower. The additional complication was the variable molecular weight of the wet gas. The molecular weight of air is constant at 29. The molecular weight of wet gas varied from 30 to 36. I noted that the performance curve shown in Figure 15-3 is flow vs. head, not flow vs. differential pressure.

What is the difference between head and differential pressure? The difference is gas density. And gas density is proportional to molecular weight. If the gas molecular weight increases by 20%, so will the differential pressure developed by the compressor. Note that the compressor discharge pressure, P_2, is constant because of the PC valve on the dry gas outlet. Therefore, the compressor suction pressure at P_1 will drop due to the increased wet gas molecular weight.

"*Pardon, Señor* Lieberman," interrupted the PDVSA technical director. "This process using the wet gas compressor cannot permit the drum pressure to fall. The pressure at P_3 must remain constant. We operate with the suction throttle valve bypassed. I control my compressor suction pressure with the spillback valve, as shown in Figure 15-4."

"And this control method is okay?" I asked.

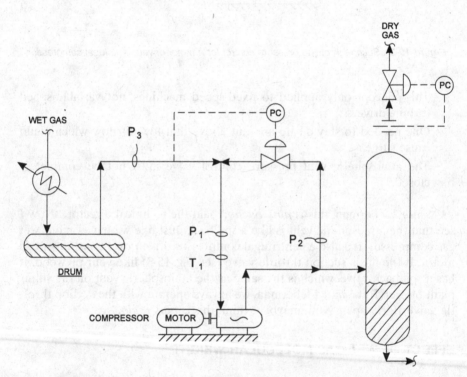

Figure 15-4 *Spillback suction pressure control. Constant-speed compressor*

"Sadly, no," replied the director. "It results in *muy mucho* power consumption. Sometimes my motor trips off on *muy alto* amperage load. *Es muy malo.*"

"That's what the PC suction throttle valve is for, *Señor Director.* It's there to keep P_3 constant (Fig. 15-3), as the molecular weight goes up," I explained.

"But *Señor* Norman, closing the suction throttle valve will raise the compression ratio and waste even more motor power and amperage."

"No, no, no," I answered. "As the suction throttle valve closes, here is what is going to happen:

- The pressure at P_3 will rise.
- The spillback FC valve will close.
- The compression ratio $(P_2 \div P_1)$ will increase.
- The number of moles of gas compressed (see *Eq. 1*) will decrease.
- Because the compression ratio is raised to a small fractional exponent, the amperage load on the motor driver will drop for a given molecular weight of wet gas."

"Ah! Just like the sulfur plant air blower. *Bueno.* But," asked the technical director, "suppose the FC spillback control valve closes too much. Maybe my compressor will surge."

"*Señor*, take another look at the controls on Figure 15-3. See," I explained, "that the spillback FC control valve is not measuring the spillback recycle flow, but the total gas flow to the compressor. This flow control valve is set to maintain a sufficiently high flow to protect the compressor from destructive surge."

"*Si, si, si. Yo comprendo.* Just like the air blower." *El Director* was now on a roll.

"*Mira tonto*," he explained to the expert engineer from the Conservation Survey Corporation, "we should start using the suction throttle valve on the wet gas compressor. We should set it to work on closed-loop automatic pressure control to control the drum pressure in Figure 15-3. We should set the FC control valve for the minimum safe flow to prevent the wet gas compressor from surging."

"But the variable molecular weight. What will happen then if. . . ." The survey corporation expert was wonderfully confused.

"No, no," exclaimed the PVDVA technical director. "This is not a problem. The suction throttle control valve PC will close to maintain a constant drum pressure at P_3. Certainly the pressure at P_1 will drop. Certainly the compression ratio, $P_2 \div P_1$, will increase. But it's *no importante* because the compression ratio is raised to a small fractional exponent. *Mira tonto*, it's just like the sulfur plant combustion air blower. It's just like the vent valve and *Señor* Lieberman's plywood."

SUMMARY

When we have a constant speed centrifugal compressor working on gas of variable molecular weight, we have two objectives:

1. To keep the compressor out of surge.
2. To minimize the amperage load on the motor driver.

To optimize those two objectives we also have to optimize the position of the suction throttle valve and the spillback valve. It's rather like solving two equations with two unknowns. We have to solve the equations simultaneously.

To arrive at the optimum solution we have to optimize the setting of the spillback control valve and the suction pressure control valve simultaneously.

In practice, this problem requires advance computer control. There are several organizations that market software programs that optimize antisurge protection and concurrently minimize driver horsepower.

Compressors that have variable speed drivers have a family of operating curves rather than the single curve shown in Figure 15-2. That is, there is a curve for each speed. The inclusion of this third variable to optimize compressor speed makes it even more essential that the centrifugal compressor be controlled by an advanced software computer program.

16

Controlling Centrifugal Pumps

The major process objectives of controlling a centrifugal pump are:

- Protect the pump from cavitation. Cavitation causes damaging vibration to the pump's mechanical seal, bearings, and impeller.
- Sustain the required process flow.
- Avoid excessively low flow rates. This also causes damage in larger pumps to the pump's internal components.
- Save electric energy or motive steam that is used to drive the centrifugal pump.
- Prevent excessive turbine driven pump speed.

This last item is discussed in Chapter 17, "Steam Turbine Control."

Pump discharge flow should normally be controlled by the pump suction pressure rather than maintaining a level in an upstream vessel. While very prevalent in the process industry, level control should normally not be used to control the discharge valve of a centrifugal pump. I explain this rather novel statement below.

PUMP SUCTION PRESSURE VS. LEVEL CONTROL

The application of level control in most process vessels is wrong. The primary purpose of many vessels is to stabilize flow rates to downstream equipment,

Troubleshooting Process Plant Control, by Norman P. Lieberman
Copyright © 2009 John Wiley & Sons, Inc.

and these vessels do require a variable inventory. We flow control out of such vessels where the flow is reset by the vessel level. I'm not concerned with such services here, but just simple level control. Simple level control has dual objectives:

1. To keep the level below the bottom reboiler return nozzle, vapor inlet, or bottom tray
2. To keep the level high enough to provide adequate suction head pressure to the downstream bottoms pump

To restate the above dual objectives, we wish to keep the level as low as possible, consistent with providing adequate NPSH (net positive suction head) to the pump. The correct way to fulfill this objective is by suction pressure control as shown in Figure 16-1. The suction pressure tap is located downstream of the suction filter. In case the vessel bottom nozzle or the filter itself plugs off, the pump discharge valve will close to protect the pump's mechanical seal. When operating on level control, this valve would open if the nozzle plugs and the mechanical seal would then be damaged because of cavitation. Level indication may still be desirable, but the pump discharge valve is typically best controlled on pump suction pressure. Best of all, the loss of pump reliability due to plugged level taps will be eliminated.

In Figure 16-1 again note that the pressure sensing point is downstream of the pump's suction screen. This is a vital point. When I first tried this concept, on a diesel product pump used at the crude distillation plant in Aruba, the result was an utter failure. The plant manager of the Coastal Refinery, Mr. English, turned red with anger. The problem, he explained, was that the suction screen plugged. Then the pump discharge valve would open 100% and the pump would cavitate. The resulting vibrations caused the mechanical seal to fail. The pump caught fire. The correct and safe location for the suction pressure transmitter is downstream of the pump's suction screen. This will require dual suction pressure transmitters, one for the spare and one for the main

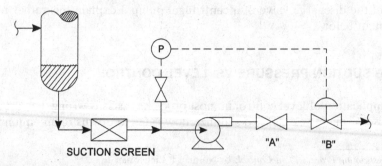

Figure 16-1 Suction pressure control with fixed-speed pump. Level control not required

pump. Also, the operators must remember to switch the transmitter output cascading to the downstream control valve when they switch over to their spare pump.

Another early failure of this novel control scheme occurred at Chevron's Refinery in El Segundo, California. The service was vacuum tower overflash. This black, heavy hydrocarbon stops flowing at ambient temperatures between 50 °F and 75 °F. Chevron connected the pressure transmitter, shown in Figure 16-1, to the pump's suction tap by 20 feet of bare, uninsulated, no steam tracing, half-inch stainless steel tubing. The connection plugged the first minute. The overflash pump over-heated and blew the mechanical seal. Why? Because the pressure at the transmitter dropped to zero and the discharge process control valve was driven 100% closed.

I close-coupled the pressure transmitter and had it steam traced and wrapped with insulation. Afterwards all went very well. Incidentally, the whole concept of suction pressure control originated with the operators at Chevron, El Segundo.

DETERMINING SUCTION SET POINT PRESSURE

This should never be done by calculation. My field-tested method described below is much better:

- **Step One**—Place a pressure gauge on the suction of the pump shown in Figure 16-1. If there is no connection to install such a gauge on the suction, use the spare pump. You can use the discharge pressure of the idle spare pump as long as it is not running and the discharge isolation gate valve is shut. This rather crude expedient is really just to demonstrate my method. In routine practice the pressure tap connection as shown in the sketch is required.
- **Step Two**—Valve "B" (the level control valve) should be switched from automatic to the manual mode of control at the control console.
- **Step Three**—Close the pump discharge isolation gate valve halfway. Station yourself at the valve and carefully watch the new pump suction gauge pressure reading.
- **Step Four**—Have the panel board operator slowly open control valve "B." By slowly, I mean a few percent of the valve positions every 3–5 minutes.
- **Step Five**—The pump suction pressure you see will not change but for 1 or 2 psi for a while. Suddenly, at some point the suction pressure will decline much faster. The liquid level has now dropped out of the vessel and into the pump's suction piping.
- **Step Six**—Quickly throttle back on valve "A" to stabilize the pump suction pressure. Your target is the pressure you observed when the pressure first started to decline rapidly on the local pressure gauge.

- **Step Seven**—Instruct the panel operator to manually open control valve "B" to 100% open on the console.
- **Step Eight**—I like to spend 10–20 minutes controlling the suction pressure by hand on the gate valve shown in Figure 16-1. I want to make sure that the pump will not cavitate over some reasonable range (i.e., 1 or 3 psi) of pump suction pressure.

For this method to work properly there are a few limitations to keep in mind. First, NPSH available must not be marginal. If a tower level needs to be rather high in the column, I'm not too sure (having never tried it) that suction pressure control will work too well.

Second, the tower pressure cannot be excessively variable. For low pressure towers, but especially for vacuum fractionators, this is not a problem. For higher-pressure towers above 40 or 50 psig the pump suction pressure control set point would have to be reset by the fractionator pressure transmitter. This is a simple computer application—I think. But to be honest, this too I have never tried on automatic control. On the other hand I've retrofitted a dozen vacuum towers and low (i.e., less than 2 bar)-pressure towers with suction pressure control, all with totally successful results.

Finally, if the vessel upstream of the centrifugal pump is actually being used for inventory surge control to dampen out flow swings to downstream equipment, suction pressure control is not applicable.

TURBINE-DRIVEN PUMPS

The above discussion is for fixed-speed, motor-driven pumps. For variable-speed, turbine-driven pumps, there is a better method, better because it is simpler and more energy efficient, as shown in Figure 16-2. Personally, I've never used this concept in a new design. But I've seen it working twice, once at an old crude unit at the Gulf Refinery in Port Arthur, Texas and once at the even older Getty Plant, Alky Unit, in Eagle Point, New Jersey. In both cases it worked so well, for so long, that the operators insisted that the pumps controlled themselves and that no control was needed! When the pump suction pressure fell below the set point, the speed control caused the turbine steam governor valve to close. The flow of the motive steam to the steam turbine shown in Figure 16-2 decreased. As the turbine driven pump slowed, the pump suction pressure rose back to its set point pressure. It's rather elegant. It's exotic, beautiful, and perfectly simple. Not only have we eliminated one control loop (the level control loop), but we have also eliminated the pump discharge control valve. By eliminating this valve we have reduced to zero the parasitic control valve losses that I describe in Chapter 10, "Sizing Control Valves."

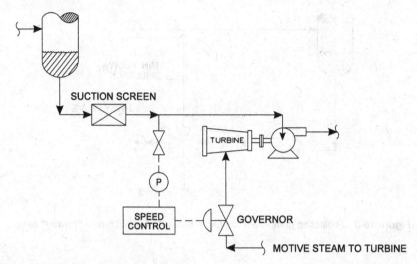

Figure 16-2 *Suction pressure control with variable-speed pump. Level control & discharge valve on pump not required*

Variable-Speed Motors

I also recall that at the Chevron Plant in El Segundo they have a giant gas oil circulation pump that has a variable-speed motor. Such pumps run at varying speeds by altering the frequency of the electric power to the motor driver. The pump circulation rate (FRC) is controlled by frequency variation rather than by wasting energy by throttling parasitically on a downstream process flow control valve. I've been told that frequency control of pump speed is becoming progressively less costly with improvements in electronics. Regrettably, I have never yet used such an electronic innovation in any process control design. But you should, and I will, too.

SAFE MINIMUM FLOW CONTROL

One control problem that may damage the mechanical seal of a centrifugal pump is running the pump at too low a rate. Approximately, for small pumps (20 horsepower or less) a pump should not be run at less than 10–20% of its design point. Very approximately, for larger pumps (200 horsepower or more) a pump should not run at less than 50–60% of its design point. Below these rates pumps will suffer from internal surge that promotes damaging vibrations.

A three-way valve, as shown in Figure 16-3, is used to prevent excessively low flows. "Yarway" is a trade name for one such three-way valve. My problem is not with the three-way valve, it is with the thermal affect of excessive spill-

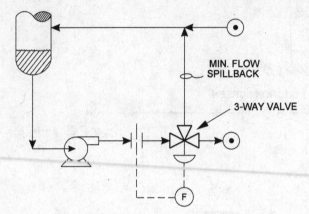

Figure 16-3   *Protecting pump from too low a flow with minimum-flow "Yarway" valve*

back flow. If most of the pump flow is recirculating through the three-way spillback valve port, then the feed vessel will overheat. I was starting up a naphtha hydrotreater unit in Aruba when I first encountered this problem. The 1200 psig pump discharge flow was mainly recirculated to the feed drum. The pumping energy was largely converted to heat in the recirculation "Yarway" valve. After two hours the feed drum had gained 50 °F. I lowered the total flow control set point, and the spillback flow decreased enough to stabilize the feed vessel temperature.

Another problem with many three-way valves is that there is a tendency to leak. The liquid leaks through to the spillback line, even though the spillback port on the Yarway valve is shut. This wastes energy and reduces the pump's capacity and head.

To calculate the amount of heat generated by a spillback, proceed as follows:

- First, calculate the head lost, in feet, through the spillback valve:

$$(PD - PV) \cdot (2.31) \div (SG) = \text{``A''}$$

where:
  PD = pump discharge, psig
  PV = vessel pressure, psig
  SG = specific gravity of liquid

- Second, calculate the weight of liquid pumped in pounds per hour.

$$(GPM)(60)(8.34)(SG) = \text{``B''}$$

where:

$$GPM = \text{U.S. gallons per minute (hot)}$$

- Third, multiply term "A" (feet) by term "B" (pounds per hour) and divide by 778. The result is the heat in BTUs per hour generated through the spillback control valve.

UNDERSIZED CONTROL VALVE REDUCES PUMP CAPACITY

One common job of the Process Control Engineer or technician is to determine whether a pump capacity limitation is because the downstream control valve is too small or if the pump itself is causing the limitation. It's rather futile to answer such a question based on the control valve characteristics or the pump curve. The pump impeller may be worn or the control valve trim can be smaller than the plant records indicate. From the perspective of the plant operator, only one question matters. Should she have the pump overhauled or repair the process control valve? To respond to this question, we should proceed as follows:

- **Step One**—Force the control valve to 100% open position, by either temporarily raising the flow or partially closing the isolation valve at the discharge of the centrifugal pump.
- **Step Two**—Open the control valve bypass valve 100%. Observe the effect on flow.
- **Step Three**—If the flow increases by a few percent, then the problem is with the pump itself. Perhaps the impeller-to-case clearance has increased. More commonly, the impeller wear ring needs to be replaced.
- **Step Four**—If the flow increases by 20% or more, then the problem is with the control valve. Perhaps the valve stem is not being pushed up to its maximum position because of a lack of instrument air pressure to the diaphragm. Or maybe the diaphragm itself is broken. It might be that the stem is stuck in a less than fully open position. I once had a pebble jammed in the valve seat. On another occasion the control valve seat was loose and vibrating. Most common of all, the control valve internal trim is smaller than the maximum size that can be accommodated in the valve body.

Energy Saving Suggestion

Let's say we have the opposite situation. That is, the process control valve is running in a mostly closed position. In the case of a turbine-driven pump we would simply reduce the turbine speed. In the case of a motor-driven pump, we would reduce the diameter of the pump's impeller. For a reduction of either the speed or the size of the impeller diameter by X%, the savings in energy would be proportional to X% raised to the third power. This is an application of the infinity or fan law, which states that work varies with speed cubed, or the diameter of the rotating element, also cubed.

17

Steam Turbine Control

The objective of controlling a steam turbine is to extract the required amount of work from the turbine using the minimum amount of steam. We have two control points that must be optimized to achieve this objective:

- The governor speed control valve
- A set of nozzles properly called horsepower valves, but commonly referred to as hand valves. I will call them hand valves or nozzle port valves.

Hand valves are never adjusted in a partly open position. To do so will degrade the valve seat by erosion. In Figure 17-1, I have shown three hand valves in parallel. This is a process sketch and does not represent the physical configuration of a steam turbine. The objective of the Process Control Engineer is to maximize the pressure in the steam chest at P_2. This will always result in minimizing the steam consumption per unit of work extracted in the turbine case. But, before I can explain how to control the turbine, to maximize the P_2 steam chest pressure, I should first explain how a turbine works.

STEAM TURBINE THEORY

Referring to Figure 17-1, the motive steam pressure at P_3 is assumed constant. Also, the exhaust steam pressure at P_1 is constant. The pressure in the turbine case at P_1 is identical to the exhaust steam pressure. It is quite wrong to think

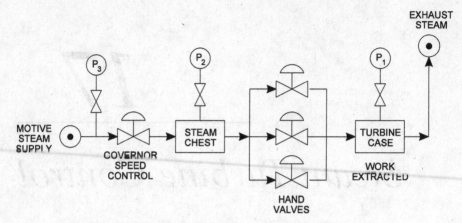

Figure 17-1 Steam turbine component functions

that it is the differential pressure of the steam, $P_3 - P_1$, that drives the turbine. Rather, the turbine works like this:

- **Step One**—High-pressure motive steam flows through the governor speed control valve. This valve maintains the turbine spinning speed at a fixed RPM (revolutions per minute). The heat content of the steam does not change as it flows through this valve. Also, the kinetic energy of the steam is assumed to be constant. As a chemical engineer I would say the enthalpy of the steam has remained constant. But our ability to extract useful work from the flowing steam has been diminished. As a chemical engineer I would say the entropy of the steam has been increased because of the pressure drop across the governor valve.

- **Step Two**—Inside the steam chest there are three outlet nozzles called ports. These nozzles are specially shaped for a particular function. That function is to convert the pressure of the steam to kinetic energy as efficiently as possible. Rather like a garden hose, you adjust the nozzle at the end of the hose to convert the water pressure to velocity. As the steam passes through the hand valves, the heat content of the steam is reduced. However, the kinetic energy of the steam has increased. The nozzles in the steam chest, or the hand valves, convert the heat content of the steam to kinetic energy. As a chemical engineer I would say that the enthalpy of the steam has been reduced, but the entropy of the steam is constant. Constant entropy means the ability of the steam to perform work has not been diminished.

- **Step Three**—The steam is now moving at a very high velocity, typically over 1000 feet per second. The speeding steam strikes the turbine blades and makes the turbine spin rather like a windmill. The steam slows down. As work is extracted from the steam it also partly condenses. Typically,

10% of the steam might turn to liquid water. The heat content of the steam is diminished as the steam's heat is converted to the rotational energy of the spinning turbine rotor.

• **Step Four**—Having done its job, the exhausted steam and condensate mixture now flows into the steam exhaust line. The greater the pressure at P_2, the greater the percentage of condensate in the exhaust steam. But this is good. More condensate means a greater percentage of the steam's enthalpy content has been converted to useful work.

I call the pressure drop through the governor speed control valve a parasitic loss, in that the ability of the motive steam to do work is reduced. I call the pressure drop through the hand valves a useful expansion, in that the heat content of the steam is converted to velocity, useful in the sense that it is the velocity of the steam striking the turbine blades that causes the turbine to spin.

USE OF THE HAND VALVES

By now you should have understood that the higher the pressure in the steam chest at P_2, the more pressure is available to convert to velocity through the hand valves. To maximize the pressure at P_2, we would close one of the three hand valves shown in Figure 17-1. This would then initiate the following sequence of events:

• The flow of steam into the turbine case would initially drop by 33%.
• The turbine would slow down by about 10%.
• A control signal would open the governor speed control valve.
• The flow of the steam that had dropped by 33% would be partly restored.
• The pressure drop across the governor valve would be reduced because the valve has opened.
• The pressure in the steam chest at P_2 would increase.
• The flow of steam entering each of the two open hand valve nozzles would increase.
• The velocity of steam exiting from the two open hand nozzles into the turbine case at P_1 would also increase.
• The kinetic energy of the steam impacting against the turbine blades would increase. That is, more work would be extracted from each pound of steam.
• The turbine would finally return to its original set speed.

As more work would be extracted from each pound of steam, less steam would be required to spin the turbine. If the pressure at P_2 was still well below the pressure at P_3, a second hand valve could be closed. Of course, if all three hand valves were shut, the turbine would stop.

The Process Control Engineer therefore has two parameters to be optimized. These are the pressure at P_2 (which should always be at a maximum) and the turbine speed (which needs to be optimized). I will describe later how to optimize the turbine speed. First, though, let me describe the steam rack.

Steam Rack

The turbine steam rack looks like a long arm that moves a row of plungers up or down. The long arm is just opening or closing the hand or horsepower valves that I have just discussed. The steam rack is opening or closing the nozzle port valves, one at a time. It's a closed-loop control that is trying to maximize the pressure at P_2, in the steam chest. Or the steam rack is trying to keep the governor speed control valve in as wide open a position as possible. Or, better yet, the long steam rack arm is trying to minimize the pressure differential between P_3 and P_2. These objectives all mean the exact same thing. That is, the steam rack is trying to avoid parasitic pressure losses across the governor speed control valve that degrade the ability of the steam to do work. If the pressure at P_2 is below the set point, then the steam rack would close one nozzle port valve in the steam chest shown in Figure 17-1. If the pressure at P_2 is above the set point, then the steam rack would open one nozzle port valve. When I said that the pressure at P_2 was above the set point, that's the same as saying the governor speed control valve is 100% open, speed control has been lost, and the turbine is slowing below its set point RPM.

To summarize, the steam rack is just performing the function automatically that I did manually with the hand valves, except that when a steam rack is used, the valves are not referred to as hand valves, but nozzle port valves.

What I have just described was the application of thermodynamics at the Amoco Refinery in Texas City during the long strike of 1980. I was working as a replacement for the striking workers at the sulfur recovery plant. More precisely, I had met a young lady at the refinery who I was trying to impress with my ability to save steam. I can't say I was very successful, so I tried another approach, that being optimizing the turbine speed.

Optimizing Turbine Speed

The speed of a turbine is variable. The operator selects the desired turbine speed, typically in the range of several thousand RPM. The amount of work needed to spin the turbine is dependent on the amount of work needed to drive the centrifugal pump shown in Figure 17-2. The turbine and pump in this simple example are directly connected with a coupling. There are no intervening gears. If the pump runs faster, the amount of work needed to drive the pump increases with the speed cubed. For example, increasing the pump speed from 2000 rpm to 2200 rpm (by 10%) would increase the horsepower needed

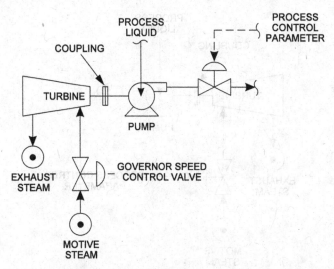

Figure 17-2 *Turbine-driven process pump*

to drive the pump from 3000 to 4000 horsepower (by 33%). The motive steam required to drive the turbine would also increase by 33%. Running the turbine and pump slower would save a lot of steam.

To select the optimum turbine speed, the operator reduces the turbine speed control point to the governor until the downstream control valve shown in Figure 17-2 is in a mostly open, but still controllable, position. For every 3% reduction in turbine and pump speed, 10% of the turbine driver steam is saved. But as far as achieving our objective of automated control, there is a much better method. It's an old idea still seen on older process units. I've only seen it three times in my long career, but it does work beautifully.

Direct Speed Control by Process Parameter

As summarized in Figure 17-3, the process control valve on the pump discharge is eliminated. Also, the operator no longer selects the turbine's desired set speed. The process parameter to be controlled (level, flow, or pressure) directly controls the motive steam flow via the governor speed control valve. In this way the turbine is always running at that minimum speed needed to satisfy the process requirements.

The parasitic energy lost across the process control valve, downstream of the pump, is not only minimized but totally eliminated. The cost of the control valve is saved, and one control loop does the job of two control loops. Why this excellent automated method of direct control of the turbine ever fell into disuse in the process industry is a real mystery.

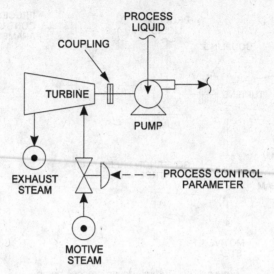

Figure 17-3 *Direct speed control by process parameter*

Well, the young lady was impressed by my combined efforts to save steam by optimizing both the hand valve positions and turbine speed. But it was all in vain. "Norm," she said, "Our problem is that we're both married, but not to each other."

Retrofitting a steam turbine to control the speed automatically on a closed loop, based on optimizing turbine speed, is a relatively simple matter. I do not believe that there is any practical method to convert a turbine with hand valves to an automated steam rack. Also, smaller steam turbines of less than 500 horsepower never come equipped with steam racks, as this type of automation is expensive. But, whenever possible, the Process Control Engineer should insist that the extra expense of the steam rack is justified based on energy conservation objectives.

One final note: Always check the delta P between the steam chest (P_2) and the motive steam (P_3). The pressure drop, when the governor is 100% open, should be less than 20–25 psi. If it is greater, then something is amiss with the governor speed control valve. Typically, the governor valve is not actually opening 100%. The governor valve position indicator may be faulty. The plant machinist will be able to correct this costly and energy inefficient malfunction.

TURBINE OVER-SPEED TRIP

One of the very first automated closed-loop controls ever developed by mankind was the steam engine over-speed trip. James Watt invented this device in the eighteenth century. Before this development, steam engines had a definite tendency to self-destruct because of excess speed. All steam turbines are protected from excessive speed by an over-speed trip. The over-speed trip theoretically serves as a backup to the governor speed control valve. A typical steam turbine runs up to 3600 rpm, with the over-speed trip completely stopping the motive steam flow at 3750 rpm (see also Chapter 22, "Alarm and Trip Design for Safe Plant Operations" and Chapter 19, "Function of the Process Control Engineer").

From what I have said, it rather seems as though we have two closed-loop control systems working in tandem, that is, the over-speed trip plus the governor speed control valve. This is certainly the conventional explanation as to how the governor and trip work together. That is, there is redundancy in the dual control loops to provide protection from over-speed. Most unfortunately, this is not quite true.

The problem arises from the characteristics of a centrifugal pump. When the process discharge valve shown in Figure 17-3 closes, the flow is reduced and the discharge pressure rises. But the flow drops off to a larger extent then the increase in discharge pressure. The work produced by the turbine to drive the pump is proportional to flow multiplied by the delta P developed by the process pump. Therefore, the required work needed to drive the pump diminishes as the discharge process control valve closes.

Now let's assume the governor valve is stuck. The steam flow to the turbine is constant. As the process valve on the pump discharge closes, the pump and the turbine run faster because less power is needed to drive the pump. At some point the over-speed trip safely and properly shuts off the steam supply to the turbine.

What this means in practice is that during normal operations of a centrifugal process pump, both the over-speed trip and the governor speed control valve must be fully functional. If the governor is not working, then the variation in load on the turbine will cause repeated over-speed trips. Operational personnel may then be tempted to disable the over-speed trip. Workers have been killed (i.e., the Coastal plant in Corpus Christi, Texas) in this manner. If the over-speed trip is not working, it is obviously unsafe to run the turbine.

In conclusion, the over-speed trip and the governor speed control valve are a team. Both members of the team are required to run the turbine safely. Hence, there is not really any redundancy. If one of the members of this team is not working, then there is simply no way to operate the turbine safely and the turbine must be shut down until repaired. I take this all personally. I informed the management at the Coastal plant about the disabled turbine over-speed trip mechanism, but they ignored my warnings.

18

Steam and Condensate Control

.

Many process plants recovery very little steam condensate. This increases the volume of plant effluent to be treated. Also, the cost of turning raw water into de-aerated boiler feed water is large. Normally the condensate loss is due to improperly designed controls for condensate drainage. The operators deal with those control problems by bypassing the controls and draining the condensate to the sewer. Improper condensate drainage control reduces the capacity of steam reboilers and heaters. Instability of the associated process equipment is also a consequence of condensate flow deficiencies.

To introduce this problem let's look at Figure 18-1. How is this control scheme supposed to work?

- **Step One**—Low-pressure steam flows through an FRC (flow recorder control) valve. The delta P through this valve will be appreciable and variable, appreciable in the sense that the delta P will be a substantial percent of the steam supply pressure and variable in the sense that as the FRC valve moves, the delta P will also change.
- **Step Two**—The variable pressure steam will condense in the tube side of the heat exchanger. The condensate drains through the outlet nozzle into the steam trap.
- **Step Three**—The steam trap is really a level controller. A ball float is lifted by the water level in the trap. This allows condensate to drain out of the trap.

Troubleshooting Process Plant Control, by Norman P. Lieberman
Copyright © 2009 John Wiley & Sons, Inc.

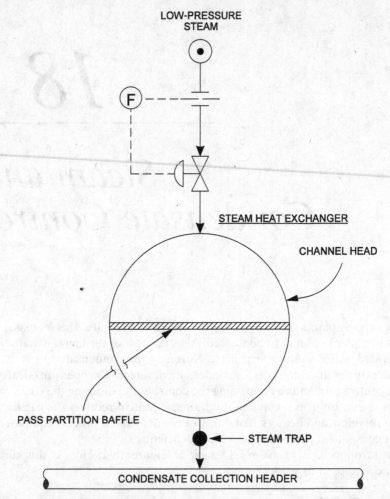

Figure 18-1   *Low-pressure condensate flows erratically into a collection header*

- **Step Four**—The condensate flows into a condensate collection header. There are hundreds of other streams also flowing into this header. Thus the pressure in the condensate collection header is intrinsically erratic.

The cause of the instability in the exchanger shown in Figure 18-1 is that the pressure in the steam trap must be greater than the pressure in the condensate collection header. If the pressure in the header is larger than the pressure in the trap, then the steam trap float will be lifted. Then condensate will flow into the channel head of the exchanger. The exchanger will fill with condensate, and the resulting water backup will reduce the steam flow into the exchanger. The FRC valve will then swing open. Steam pressure in the channel head will rapidly blow the condensate out of the channel head. The sudden flow of steam will overheat the shell-side product. Then the FRC valve will close. The channel head pressure will fall, and the cycle will be repeated.

The panel board operator will not tolerate this erratic heat exchanger duty. To stop the instability he directs the outside operator to divert the steam trap condensate effluent to the sewer.

CONDENSATE LEVEL CONTROL

Figure 18-2 shows an acceptable control strategy to deal with the condensate backup problem. The pressure in the channel head is now constant at the full

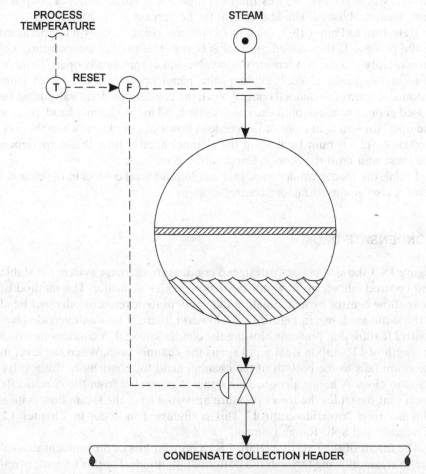

Figure 18-2 *Condensate level controls flow of steam into the exchanger but avoids blowing the condensate*

pressure of the supply steam. If too much steam flows into the exchanger, the condensate drain valve closes. The condensate backs up. This reduces steam flow in a controllable fashion. The difficulty arises when the level drops too low. This occurs because the panel operator cannot see the water level in the channel head. Suddenly, and unexpectedly, the water drains totally out of the channel head. Steam begins to blow through the outlet nozzle. This is called "blowing the condensate seal." Steam now begins to condense in an erratically low manner. The heat input to the process also becomes erratically low. The disadvantage of this control scheme is its positive feedback nature. That is, opening the condensate drain valve normally increases the heat input to the process. At some point, which is the point at which the condensate seal is blown, a further opening of the drain valve reduces heat input to the process. As the process is requiring more heat, the condensate drain valve opens even further, which further reduces the heat input. A positive feedback loop has now been established. The loss of heat to the process feeds upon itself.

Referring to Figure 18-2, the flow of steam is being reset by a temperature in the process. If the condensate seal is blown, this process temperature will drop rapidly and the condensate drain valve will automatically open. To intervene in this positive feedback loop, the panel operator must switch from automatic control to manual control Next, the condensate drain valve must be closed enough to reestablish the condensate level in the channel head. But for the operator sitting in front of the console, how does she know when the condensate level has built back up in the channel head? Other than experience, she must wait until the process temperature goes back up.

I think the reader would agree that condensate level control in the channel head is also an unsatisfactory control strategy.

CONDENSATE DRUM

Figure 18-3 shows a properly designed condensate drainage system for stable and controllable steam flow into a steam reboiler or heater. The method of steam flow control by manipulating the condensate level in the channel head is the same as shown in Figure 18-2. However, there is now an override level control feature that prevents blowing the condensate seal. A condensate drum (a length of 12″ carbon steel pipe) spans the channel head. When the level in the drum falls to the bottom of the channel head the condensate drain valve starts to close. A higher pressure air signal is generated from the condensate drum that overrides the lower pressure air signal from the steam flow orifice. This is called "override control." This is discussed in detail in Chapter 12, "Override and Split-Range Control."

The intent of the override control is to prevent loss of the condensate seal due to steam blowing through the bottom drain nozzle. For this to work properly, the level in the condensate drum must match the level in the channel head. For these two levels to match, the pressure in the condensate drum must

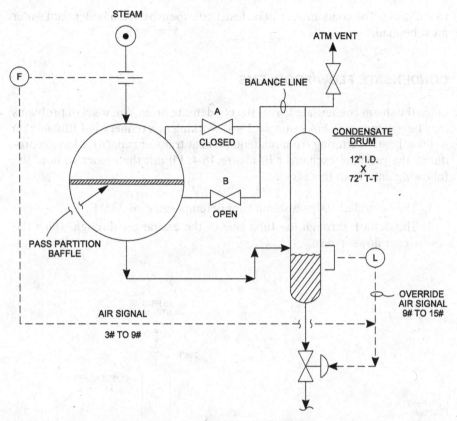

Figure 18-3 *Properly instrumented condensate level control for stability. Note override air signal*

match the pressure below the pass partition baffle in the channel head. For these two pressures to coincide, we need the balance line shown in Figure 18-3. Too often, the balance line is connected above the pass partition baffle. This is wrong because the pressure in the condensate drum will be too high.

If there is a 5 psi pressure drop of steam in the exchanger, then the pressure above the baffle will be 5 psi greater than below this baffle. The resulting high condensate drum pressure will push up the level in the bottom of the channel head and flood the exchanger tubes. A connection below the pass partition baffle for the balance line is needed for proper control and heat transfer stability. For venting of air on start-up, the connection above the baffle is required. However, as shown in Figure 18-3, valve A must be kept shut while valve B must be kept open.

If there is no connection below the pass partition baffle, then the control of steam flow to the exchanger will be very poor. To restore controllability, the panel board operator will direct the outside operator to drain the channel head to the sewer. The condensate will be lost, and expensive fresh boiler feed water must be made.

CONDENSATE FLOW PROBLEMS

Once the steam condensate leaves the condensate drum, a new set of problems may be encountered. For example, I was working in a refinery in Lithuania on a steam heater suffering from inadequate heat transfer capacity. I have reproduced the problem exchanger in Figure 18-4. I'll ask the reader to note the following data from this sketch:

1. The saturated 100 psig steam supply temperature is 335°F.
2. The delta P through the tube side of the exchanger through which the steam flows is zero.

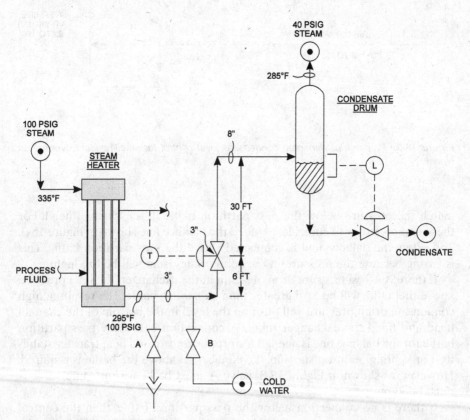

Figure 18-4 *Flashing condensate causes backup in steam heater*

3. The condensate effluent leaving the exchanger is only 295 °F. This means the steam condensate leaving the exchanger is subcooled by 40 °F below its boiling point temperature of 335 °F.

4. The pressure loss for the flowing condensate between the heat exchanger outlet and the inlet to the condensate drum is 60 psi (100 psig – 40 psig).

5. The inlet to the condensate drum is 36 feet above the steam condensate outlet nozzle.

6. The line size before the control valve is 3″. The line size downstream of the control valve is 8″.

7. The control valve draining the heat exchanger itself is 3″.

8. The control valve is elevated by 6 feet above the exchanger condensate outlet nozzle.

What does this have to do with control engineering? It depends on how one defines the Process Control Engineer's responsibilities. I believe these include the selection of the location and elevation of control valves. How could an improper location of the condensate outlet temperature control valve cause lack of steam heat exchanger capacity?

Here is what happens:

- **Step One**—At time zero the saturated condensate leaves the exchanger at 325 °F.

- **Step Two**—The condensate flows upwards to the elevated control valve. As a liquid flows to a higher elevation, the liquid loses head pressure. In this case, about 2-1/2 psi is lost.

- **Step Three**—As the saturated condensate loses pressure it starts to partly vaporize into steam. Only a few weight percent of the condensate vaporizes. However, a pound of steam occupies about 300 times as much volume at these conditions, as does a pound of water. Therefore, the volume of fluid flowing to the control valve greatly expands.

- **Step Four**—The Process Control Engineer sized the condensate drain temperature control valve for water, not for steam plus water. The valve simply does not have sufficient capacity to handle a large volume of steam without an excessive pressure drop.

- **Step Five**—We could say that the temperature control valve chokes the flow. As a result, the condensate level in the heat exchanger is backed up. Condensate fills most of the length of the tubes. Inside the tubes the condensate is subcooled by the cooler process fluid on the shell side.

- **Step Six**—The condensate is subcooled below its boiling point. It is subcooled enough so that when it rises by 6 feet and loses even more pressure as it passes through the control valve, the condensate is sufficiently subcooled so that it does not flash.

If the condensate is not subcooled enough and does partly vaporize, then the condensate will continue to back up and become progressively colder until all vaporization in the 3″ line and the 3″ control valve is suppressed. Once the steam condensate enters the section of 8″ line shown in Figure 18-4, flashing of condensate to steam has little effect. The larger 8″ line had been sized correctly to accommodate the several weight percent of steam generation without excessive line pressure drop.

To stop the loss of the steam heater capacity, the operators must open valve A, which puts condensate to the sewer. This was a particularly bad problem in the Lithuanian refinery, which was very limited in boiler feed water (BFW) production.

To solve this problem in the short term, I connected a hose of cold water to drain valve B. The cold water reduced the temperature of condensate upstream of the 6-foot riser and the restrictive temperature control valve. This stopped the condensate from vaporizing. The 295 °F condensate drainage temperature shown on Figure 18-4 rose to 325 °F. This indicated that the condensate level in the steam heater tubes was falling. Gradually the exchanger process line outlet temperature rose above its set point. I increased the flow of cold water a bit more, and the temperature control valve began to close.

The cold water had restored the stability and controllability for heating the process fluid without having to drain most of the steam condensate to the sewers of Lithuania. Unfortunately, the mixing of the cold plant water ruined the condensate for reuse as boiler feed water because of the hardness deposits in the cold water. Thus a permanent solution was required.

The temperature control valve was relocated to a lower elevation. Also, it was resized to a larger valve so that when 100% open it had an acceptable pressure drop. The 8″ condensate line was then extended by 6 feet down to the new control valve station.

For me, this was a typical process control problem. An improper control valve elevation had resulted in a loss of control of a key process variable. I believe the reader can see why the university professors in Chapters 1 and 2 became angry with me. It wasn't my teaching methods that they found so objectionable. It was the very nature of the subject to be taught that I had questioned. Should Process Control Engineers be instructed in the application of state-of-the-art technology in control theory? Or is the technology of the 1920s all that needs to be mastered?

I've spent a rather long time explaining the apparently simple subject of draining hot water from a heat exchanger. I have tried to make a point. Hidden in simple subjects are often problems of great engineering complexity and importance.

19

Function of the Process Control Engineer

I woke last night to the sound of thunder. "How far off?" I thought and wondered.

The lightning flashed through my room. I waited for another peal of thunder and recalled an incident from 1975. I was late for work. As I rushed through my alkylation unit to the control room I noticed that the compressor turbine trip was unlatched. The steam turbine was spinning along quite merrily; but in a tripped condition. I describe how turbine trips function in Chapter 17, "Steam Turbine Control." Also in Chapter 17 I describe why an over-speed trip was intentionally disabled by the field operator. My alkylation unit turbine trip had functioned in a normal way. It was unlatched because something in the compressor control logic had attempted to shut down the turbine and compressor. Yet the 6000 horsepower machine was still vibrating gently and spinning along happily at 4400 rpm. A few thoughts also spun through my mind:

- How could the steam flow continue to the turbine if the motive steam trip valve was unlatched? A powerful spring should have pulled the trip valve shut.
- What had caused the trip to unlatch? Had the trip unlatched accidentally, or was there really a serious problem that I ought to be concerned about?

Troubleshooting Process Plant Control, by Norman P. Lieberman
Copyright © 2009 John Wiley & Sons, Inc.

• If the compressor and turbine self-destructed because of over-speed or high vibrations, how would that affect my career and promotional path at AMOCO?

PROCESS CONTROL ENGINEER'S SAFETY RESPONSIBILITY

The most ubiquitous of the myriads of safety slogans is "Safety Is Everyone's Job." I believe this also applies to the Process Control Engineer, who exercises a critical degree of influence over plant safety through the design and operation of alarms and trips. Alarms represent a process unit's first line of defense to prevent an accident. Trips represent the last line of defense. The two main ideas of keeping alarms and trips in a safe mode of operation are redundancy and periodic testing. My steam turbine difficulties in Texas City illustrated both problems.

The East Plant Control Engineer, Bob Allen, had been assigned by the technical manager to oversee a program to field test alarms and trips. But Mr. Allen objected on principle to this assignment. His interests were in advanced, multivariable, interactive computer control. Thus individual process supervisors such as myself were left to our own resources to maintain the operational integrity of our unit safety systems. After many years I'll be meeting Bob Allen again today. We had both been retained by Stauffer Chemicals to review the P&IDs (process and instrumentation diagrams) for a new spent sulfuric acid regeneration plant being built along the Houston ship channel. Bob has been retained to consult on the process controls and I have been similarly retained to review the process equipment for operational safety. I don't imagine that Bob Allen will even remember me. It's been 33 years since our last meeting in Texas City.

LATER THAT EVENING

Unbelievable! Bob Allen and I got into a violent argument. Not so much about the new acid plant in Houston. Our argument was about that terrible incident at my alkylation unit in Texas City three decades ago. During the lunch break I reminded Bob in a friendly way about the time the turbine trip had become unlatched but the machine continued to spin.

"I remember you asking me for some help, Norm, back in Texas City in 1975 when we worked for AMOCO. We have the same potential problem now on the sulfuric acid plant booster blower. It's basically an 1800 horsepower, turbine-driven centrifugal compressor. I've made a list of the process parameters that we have to make sure will trigger the steam turbine trip valve to shut off:

• Excessive vibration to either the blower or turbine
• Excessive speed

• Low lube oil pressure to either the blower or the turbine drive
• Operator-activated emergency shutdown

"It's extremely important," Bob continued, "that we be able to test each control loop for complete integrity. You know, Norm, A SAFETY DEVICE THAT IS NOT TESTED ROUTINELY WILL NEVER FUNCTION CORRECTLY IN AN EMERGENCY."

"Yes, Bob, I quite agree. My turbine steam trip valve in Texas City was frozen open. Hardness deposits from poor-quality steam (see Chapter 20, "Steam Quality and Moisture Content") had jammed the trip mechanism. One should exercise the trip mechanism to break off these salt deposits. That's why my trip unlatched, but the spring-loaded trip valve didn't move. The problem that caused the trip to unlatch was low lube oil pressure to the compressor inboard radial bearing housing, potentially, a catastrophic problem." I concluded.

"I also remember that," said Bob. "But Norm, I've forgotten why the backup lube oil pump did not switch on automatically to sustain the proper lube oil pressure. The electronic switch to activate the backup pump should have sensed the low lube oil pressure at the radial bearing housing. Here, give me your napkin and I'll sketch out what I mean" (Fig. 19-1).

"Actually, Bob, I remember the entire incident very well. I really appreciate your sketch. I would have appreciated it a lot more in 1975 when I had the problem and didn't understand what to do next. Do you happen to recall my question about why the turbine-driven lube oil pump was run and the motor-driven lube oil pump was kept as a backup? More importantly, do you remember how worried I was as to how to test the automatic feature of the lube oil

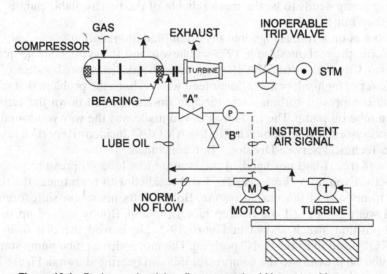

Figure 19-1 *Backup on low lube oil pressure should be motor-driven pump*

system to determine if it would actually function in an emergency. That's when I needed your sketches. Thirty-three years ago. Now I understand it all by myself."

"That was all so long ago. I was young and dedicated to advanced, multivariable, integrated, real-time computer control. I had no time for your more mundane matters. I simply was too busy to help you."

"And now Bob?"

"And now Norm, my real interest is still in advanced control techniques, which you still do not care to appreciate. However, as a process control consultant, I am required by my clients, such as Stauffer Chemicals, to advise on these mundane control problems. I am forced to work at several levels beneath my true abilities."

"Meaning my level," I said to Bob.

"Yes, Norm. Here's what you should have done in 1975. Let's refer again to Figure 19-1 that I've drawn on the napkin. First of all, the motor is a backup for the turbine because it's more reliable. The less reliable turbine-driven lube oil pump may run slow for a wide variety of process problems:

- Low-pressure motive steam
- Fouled turbine blades
- Malfunction of the governor
- High steam exhaust pressure
- Loss of steam superheat

The slow speed will produce a low lube oil pump discharge pressure. The motor-driven lube pump will run at 3600 rpm or not run at all. Clearly, the backup pump needs to be the more reliable of the two available pumps."

"Okay, but I...."

Bob was on a roll. He ignored my attempted interruption.

"Norm, the real question in 1975 was how to test the system. Steam quality at Texas City was awful. The 160 psig steam header was wet because of the carryover of high-salt content boiler feed water. That's the problem that salted up the compressor turbine's trip mechanism and slowed down the turbine-driven lube oil pump. The real mistake you made was the way you tested the lube oil system. I read about that in the AMOCO Incident Report circulated in the Technical Service Division," Bob concluded.

It was true. I had not handled the bearing low lube oil pressure problem correctly. Even after three decades, I still recalled with resentment the reference to my test of this cursed system. Bobby Felts, my senior shift foreman, and I went out to test the backup lube oil pump. Bobby walked up to the three-position switch shown in Figure 19-2. He moved the dial from the "AUTO" position to the "ON" position. The motor-driven lube pump started. The lube oil pressure to the compressor inboard bearing shown in Figure 19-1 increased by 20 psi.

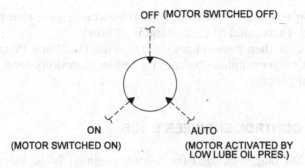

OFF (MOTOR SWITCHED OFF)

ON
(MOTOR SWITCHED ON)

AUTO
(MOTOR ACTIVATED BY
LOW LUBE OIL PRES.)

Figure 19-2 *Three-position switch governs control of the backup*

"See, Norm," said Bobby Felts, "it's all okay. The pump works fine."

"What you should have done, Norm," Bob Allen explained, "to test the system correctly, was to:

• **Step One**—Referring to Figure 19-1, partly close valve "A."

• **Step Two**—Bleed off some lube oil pressure by opening valve 'B,' which drains lube oil back to the reservoir.

• **Step Three**—The set point control pressure to activate the motor-driven lube oil pump was 40 psig. If the pressure at the bearings fell to 39 psig, the motor driven pump should have started automatically."

Bob Allen continued his long explanation, which I did not want to hear in the worst way. "When Bobby Felts just turned the three-position switch to "ON," that only proved that the motor driven pump would operate. It did not test that the pump would start up in automatic on low lube pressure to the bearings. Frank Citek, your boss, was quite surprised that such an error could have been made."

"Yes, Bob, I did make an error. There was an electrical fault in the automatic start circuit that prevented the motor lube oil pump from starting because of low lube oil pressure. I haven't forgotten."

"Too bad, Norm," Bob continued regretfully, "that the turbine-driven lube pump continued to slow down because of salt accumulation on its rotor. According to the AMOCO Incident Report that was circulated, the lube oil pressure to the bearings became so low that the bearings seized. As I recall, the resulting vibrations damaged the labyrinth seals in the compressor case. The alkylation unit was idled for several weeks to repair the damage."

"Bob, your memory is truly amazing. Maybe you can also remember that I was demoted to a junior engineering position as a result of the Incident Report, which was carbon copied to Dr. Horner, the Vice President. Maybe if you had helped me understand the concept of automatic closed-loop startup of the backup lube oil pump and how to test it online, the entire fiasco could have been avoided."

"What is it you don't understand?" Bob answered, somewhat puzzled.

"Nothing," I screamed, "I understand it all now!"

"Oh, but back then it wasn't my job. I was the East Plant Process Control Engineer. My responsibilities were on a rather more lofty level," Bob concluded rather proudly.

SCOPE OF CONTROL ENGINEER'S JOB

Experience has taught me that the control engineer is the most important component of the technical staff in an operating plant. This is not always appreciated because of an overly narrow scope of the control engineer's responsibilities.

I have in mind an incident that happened at the Texaco plant in Convent, Louisiana. The process unit involved was the propylene-propane splitter shown in Figure 19-3. I was working under contract for Wayne Hiller, the Technical Director of the refinery. Wayne asked me to review the control system for the splitter. A new, larger-capacity tower was being designed by the Texaco Engineering Division. Wayne wanted to know if the existing control logic should be duplicated in the new design, or did I have some better, more advanced concepts to recommend. Wayne explained that the existing tower was too small for current throughput requirements. Thus design and construction for the new tower was being fast-tracked. The new splitter's concrete foundation was already poured. The existing splitter had a capacity of 8800 BSD. A 10,500 BSD capacity was needed as soon as possible. Hence, the rush for my input.

Interaction With Panel Operator

I began this assignment by interviewing Jay, the panel board operator.

"Jay, how are your controls working? Do they produce stability in a reasonable time frame?"

"Norm, the controls are okay. But I've a problem with the reflux. I can't increase it above 40,000 BSD. The computer won't let me."

"But Jay, why do you want more reflux?"

"Just to run more feed," Jay responded. "If I could crank up my reflux I could increase splitter feed above the current 8800 BSD limit. You know, I've got lots of P-Ps to run off."

"Jay, don't you know why you can't increase the reflux flow? Didn't you ever ask Henry Derwinski, the West Plant Control Engineer?"

"Yeah, Henry's the one who set it up this way. If we increased the reflux rate above 40,000 BSD we lost the reflux drum level and cavitated the reflux pump. So Henry put a computer stop at 40,000 BSD of reflux to protect the reflux pump from cavitation. You know, Norm, to protect the pump's mechanical seal from vibration and damage."

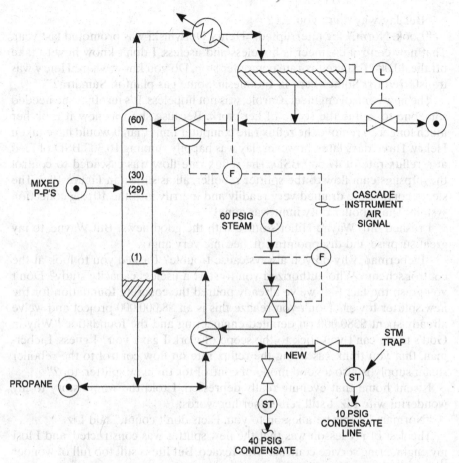

Figure 19-3 *Forgotten control limit cost Texaco $5,000,000*

"That doesn't make much sense, Jay," I responded. "Why not just increase reboiler duty to keep the reflux drum level from dropping?"

"Oh, back then we were limited on the 60 psig steam flow to the reboiler. You see," Jay explained, "the steam condensate used to drain to the 40 psig steam condensate line.

There was too much back pressure from the 40 psig condensate system. So we were limited to 52,000 lbs/h of reboiler steam. That limited the reflux rate to 40,000 BSD."

"Back then, Jay I....?"

"Well, last month process engineering and construction finally tied the condensate drain into the 10 psig condensate collection line (see Fig. 19-3). Now I can use lots more 40 psig steam to the splitter reboiler."

"But Jay, why don't you. . . . ?"

"Look, Norm," Jay interrupted. "Henry Derwinski was promoted last year. That new control engineer is hopeless and useless. I don't know how to take off the 40,000 BSD reflux rate computer stop. Do you know where Henry was transferred to? Somewhere in Indonesia. Some gas plant in Sumatra?"

The new control engineer, Carole, was not hopeless. It's just that she needed someone to define the scope of her work. Because she was new, it took her much longer to remove the reflux rate computer limit than it would have taken Henry. Three days later, however, Jay was happily running 10,300 BSD of feed at a reflux rate of 47,000 BSD. The reflux rate flow was cascaded to control the 40 psig steam flow to the splitter reboiler, all as shown in Figure 19-3. The steam condensate drained very readily and merrily into the 10 psig collection system. The reboiler duty limit was gone!

I rushed into Wayne Hiller's office with the good news! But Wayne, to my great surprise and disappointment, became very angry.

"Lieberman, why do you always cause trouble? I asked you to look at the control scheme. Who authorized you to start a useless capacity study? Don't you grasp the fact that we've already poured the concrete foundation for the new splitter tower? Don't you realize this is an $8,000,000 project and we've already spent $380,000 on detailed engineering and the foundation? Why in God's name can't you stick to the scope of work I gave you? I guess, Lieberman, that you think cascading the reflux rate on flow control to the reboiler steam supply is also a good mode of control for the new splitter, too?"

I went home that evening really depressed. I told the whole story to my wonderful wife Liz. I still remember her words:

"Norman, let this be a lesson to you. Facts don't count," said Liz.

The rest of my lesson was that the new splitter was constructed, and I lost my engineering service contract with Texaco. But life is still too full of wonder and sunlight to take such lessons too seriously.

20

Steam Quality and Moisture Content

Process control engineering is an effective venue for the application of thermodynamics. The control of the moisture content (i.e., quality) of flowing steam is an excellent example of how thermodynamics and control engineering should be employed together.

In particular. I will be describing the use of entropy and enthalpy to overcome high-moisture content problems of the steam used in refinery and petrochemical plant operations. Some of the concepts described are new and novel to the industry. However, I would not suggest something unless I had tried it myself. This sometimes means that I've done it on manual, but it may not have been converted to closed-loop automatic control. I've provided a Mollier diagram for reference, but it's best to reference your own steam tables for convenience.

FLOWING STEAM

Poor-quality steam refers to a high moisture content. Steam is best when superheated. From the prospective of the Process Design Engineer, we assume that even saturated steam is dry. In reality, steam in most process plant piping systems is wet. Often steam is wet because of ambient heat loss. For example, 150 psig steam superheated to 400 °F has only to lose 2% of its heat content to become wet.

Wet steam is generated from boilers because of entrainment of boiler feed water into the evolved steam. Entrained boiler feed water contains salts. The TDS (total dissolved solids) of the entrained water is the same as the boiler blow down (i.e., waste water drained periodically from the boiler). The salt content of the blow down water is 10 to 20 times greater than the salt content of the boiler feed water. That's why moisture in steam due to entrainment is more serious than moisture in steam due to condensation. Condensed moisture is free of salts.

But why is wet steam so bad, and what can the process design engineer do about it? Dry steam is actually invisible. Steam venting from a line only looks white because the steam is wet. Let me tabulate the consequences of poor-quality steam that I have experienced.

Vacuum Jets

A little moisture in steam will cause a vacuum ejector to surge. The jet will lose its sonic boost (see Chapter 13, "Vacuum System Pressure Control"). The operators say the jet has broken. The vendors say the jet has been forced out of its critical mode of operation. The jet will cycle between a soft, low-pitch noise and a louder, higher-pitch sound. At the Citgo refinery in Corpus Christi I noted that substituting dry steam for wet steam improved vacuum from 30 mmHg to less than 20 mmHg (millimeters of mercury; 760 mmHg is atmospheric pressure).

Wet steam, as it flashes through the steam nozzle, can cause the steam nozzle to freeze and temporarily stop the flow of steam. I saw this at the Coastal Asphalt Plant in Mobile, Alabama. The body of the jet cycled between 32 °F and 60 °F. Also, wet steam, over a period of years, erodes the steam nozzle, resulting in not only a loss of vacuum but also a waste of steam.

Hydrocarbon Steam Stripping

Water in steam will evaporate when it contacts hot oil. This cools the oil and renders the stripping steam less effective because the colder oil has a lower vapor pressure. On the other hand, stripping water or reboiling a product is not adversely affected by moisture in the steam supply.

I once corrected a low-flash diesel problem at the Coastal refinery at Eagle Point by putting a steam trap on the stripping steam supply line.

Catalyst Stripping

At the Valero Refinery in Delaware City the fluid catalytic cracking unit had a plugged steam distributor on the spent catalyst stripper. Moisture in the steam had mixed with the fluidized catalyst and turned the wet catalyst into something like cement. The wet steam was generated from a nearby kettle waste heat boiler suffering from boiler feed water level control problems.

Steam Turbine

Moisture in the supply steam contains salts. The salts slowly accumulate on the turbine blades and reduce horsepower output. When these deposits break off, the turbine rotor is unbalanced. The resulting vibration will cause a shutdown of the turbine. This problem is not common, it's universal!

Then there's the time that I came to work in Texas City and found my 5000 horsepower refrigeration compressor running with the turbine over-speed trip valve unlatched. The trip valve itself was struck in an open position because of salting of the mechanism. My boss, Frank Citek, found out about the stuck trip. He noted that this was an indication of my general incompetence (see Chapter 19, "Function of the Process Control Engineer").

Steam Superheat Furnace Tubes

The salts in entrained moisture from a boiler will deposit inside the superheat coils. Localized overheating and tube rupture will result. This happened at the Spanish Muskiz Refinery on a crude unit I had revamped. During the subsequent start-up a slug of water was introduced with the wet stripping steam and disrupted every tray in the column.

Hydrogen Plant Reforming Catalyst

The production of hydrogen involves hydrocarbon steam reforming inside furnace tubes filled with catalyst. The reforming steam is produced from a waste heat boiler. At the Coastal Refinery in Aruba, level control problems promoted water carryover into the 600 psig steam generated in a waste heat boiler. The steam dried out in the superheat coils. But the residual salts accumulated inside the reforming reaction tubes. The catalyst plugged and shut down the hydrogen plant. To overcome this problem Coastal built a third hydrogen plant in Aruba so that they always had two plants online. But perhaps there was a more cost-effective design solution to this catalyst plugging problem?

WHY BOILERS CARRY OVER

Steam has been used to drive pumps since the eighteenth century. You would have thought that by the twenty-first century the problem of controlling the water level in a boiler would have been solved. But what is the problem?

Figure 20-1 shows the 600 psig waste heat boiler in the Aruba H_2 generation plant. It's a standard Exxon design. The dual objectives of the boiler level control are to prevent cavitation of the boiler feed water circulation pumps due to lack of adequate net positive suction pressure (i.e., NPSH) and to minimize entrainment of water.

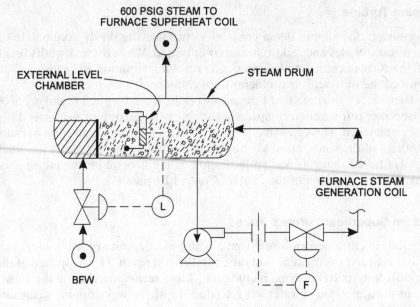

Figure 20-1   *Hydrogen plant waste heat boiler in Aruba*

Pump cavitation will destroy the pump's mechanical seal. Entrained water has 10 to 20 times the salt concentration of the boiler feed water. These salts will damage the steam superheat tubes and plug the hydrogen reformer catalyst tubes.

I've shown in Figure 20-1 that the level in the steam drum is higher than the water level in the external level chamber. The boiling water in the steam drum is less dense than the water in the external level chamber. The boiling water requires a greater height to develop the same head pressure as the still water in the external level chamber. As the water boils more vigorously, and as solids (dissolved or particulates) in the steam drum increase, the density of the boiling water drops. The level in the drum increases relative to the water in the level chamber. The ratio of these two levels is unknown and variable, because the mixed-phase density in the steam drum is unknown. This is the problem left over from eighteenth-century England.

Indirect Level Control

I did solve this problem that caused the catalyst tube plugging in Aruba. To do this I devised an indirect way to measure the moisture content or quality of the 600 psig steam flowing from the steam drum. To understand this method let's refer to the Mollier diagram (Fig. 20-2). There are two ways to expand steam—isoenthalpic or isoentropic. In both of these expansions the energy of the steam is preserved.

A MOLLIER CHART FOR STEAM

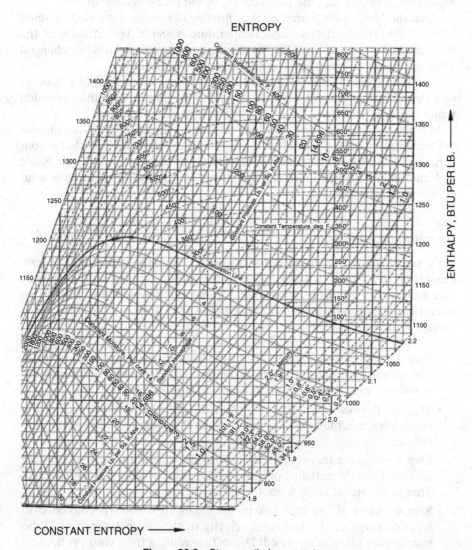

Figure 20-2　*Steam enthalpy vs. entropy*

Steam's energy appears in two forms: heat (enthalpy) and speed (kinetic energy).

Isoenthalpic Expansion

Let's say that 600 psi steam flows through a 0.35-inch restriction type orifice into a 2-inch pipe vented to the atmosphere. The ratio of the areas is such that there is no change in the velocity of the steam. This is called an isoenthalpic expansion, meaning that the heat content of the steam is constant.

From the Mollier diagram read the enthalpy of 600 psia saturated steam at 1203 BTU/lb. The saturated steam temperature is 486 °F. The enthalpy of the steam at 14.7 psi is still 1203 BTU/lb. As the steam's velocity has not changed when it enters the 2-inch pipe, its heat energy is constant.

Now follow the horizontal line at 1203 BTU/lb to the right until you intersect the 14.7 psia constant pressure line. The temperature of the expanded steam is 320 °F.

Most gases behave in this way. They cool upon expansion. In our example, the temperature drop from 486 °F to 320 °F represents 91 BTU/lb that is converted into a bigger latent heat at the lower pressure steam. Or the water molecules move further apart, and the energy to do this comes from the temperature of the steam.

Isoentropic Expansion

Let's say that 600 psi steam flows through a smooth, streamlined nozzle exhausting at 14.7 psia. It's like adjusting the nozzle on your garden hose to maximize the speed of the water escaping from the nozzles. This is called an isoentropic expansion because the entropy of the steam is constant. Let me explain with an example:

- **Step 1**—The pressure upstream of the nozzle is 600 psia and the pressure downstream of the nozzle is 14.7 psia. That is, the motive pressure is 600 psia and the exhaust steam pressure is 14.7 psi.
- **Step 2**—Because the exhaust pressure is less than half of the motive steam pressure, the nozzle velocity will be the sonic velocity of saturated 600 psia.
- **Step 3**—Assume that the nozzle is shaped for minimum turbulence and friction. Then the enthalpy or heat content of the motive steam is reduced. That is, the steam cools from 486 °F to 212 °F.
- **Step 4**—Most of the heat lost by the steam as it cools is converted to kinetic energy or the increased velocity of the steam jetting out of the nozzle. Don't be too surprised! Doesn't air venting from your car tire feel colder than the tire?
- **Step 5**—The momentum of the steam (mass times velocity) hits the steam turbine blades. The momentum of the steam causes the turbine blades to spin. This reduces the energy of the steam and transfers this energy in the form of rotational work to the spinning turbine rotor.

Let's quantify what I've just said. From the Mollier diagram read the enthalpy of 600 psia saturated steam at 1203 BTU/lb. Since the entropy of the expanding steam is constant, we will be converting the heat in the steam to kinetic energy. Therefore, follow the chart straight down until you intersect the 14.7 psia constant pressure line. What's the temperature of the expanded steam now? Well, the steam is now below the "saturation line" on the Mollier diagram. So the temperature of the expanded steam is 212 °F. But wait. Not only has the steam cooled by 274 °F (from 486 °F to 212 °F), but 22% of the steam has condensed to water. From the vertical axis of the Mollier diagram read that the heat content or enthalpy of the steam has dropped from 1203 BTU/lb to 938 BTU/lb. That's a heat loss of 265 BTU/lb. Where did that energy go to?

The 265 BTU/lb of reduced enthalpy has been converted to work, meaning that the high-velocity steam rushing out of the nozzle has transferred its force or momentum to a spinning wheel. Like the turbine rotor. As there are 2460 BTUs in each horsepower, the 265 BTU/lb worth of heat or enthalpy extracted from steam can generate 10% of a horsepower.

Calculating the Moisture Content of Saturated Steam

Hopefully, the concept of an isoentropic expansion is clearer now. So let's get back to Aruba and their hydrogen plant problem. What sort of variables can I monitor that will indirectly measure the boiling water level in the steam drum?

When I described the isoenthalpic expansion of dry 600 psia saturated steam, I noted that it would cool to 320 °F on expansion to atmospheric pressure. The saturation temperature of atmospheric pressure steam is 212 °F. The atmospheric pressure steam therefore contains 108 °F worth of superheat thermal energy (320 °F minus 212 °F). But suppose the atmospheric steam temperature observed is not 320 °F, but 220 °F. The steam is missing 100 °F or about 55 BTU/lb worth of heat. Where is this heat?

The missing enthalpy or heat has gone into vaporization. Apparently, I was mistaken when I told you the 600 psia steam was dry. It actually contained about 5.8 wt.% water, as calculated below:

$$\text{Specific heat steam} = 0.55\,\text{BTU/lb/°F}$$

$$\text{Latent heat steam} = 980\,\text{BTU/lb}$$

$$\text{Thus }(100\,°F)\cdot(0.55)\div 980 = 5.8\,\text{wt.\%}$$

Measuring Moisture in Steam in Aruba

I installed the connections shown in Figure 20-3. As I manually lowered the drum level, the new TI point temperature increased from 220 °F to 320 °F. If I persisted in lowering the drum level, the pump discharge pressure became erratically low because of lack of adequate NPSH (net positive suction head).

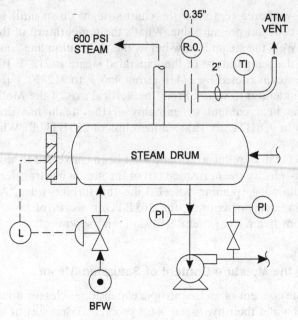

Figure 20-3 TI and PI used to determine level set point

At 70% of the indicated level, the TI point decreased. At 10% of the indicated level, the pump discharge pressure became erratically low.

I split the difference and set the level control to hold 40% of the indicated level. Neither the 10% level cavitation limit nor the 70% level entrainment limit was permanent. Both parameters changed with time and operating conditions. From the process designer's perspective what I accomplished manually could be automated. Thus the optimum LRC set point could be reset by closed-loop computer control.

LEVEL CONTROL IN A KETTLE WASTE HEAT BOILER

I must have completed 100 designs in the past 44 years that included a kettle waste heat boiler such as shown in Figure 20-4. Never once did I think about the discrepancy in level between the boiling water in the kettle versus the stagnant water in the Level-Trol. Every time I boil pasta, I've noticed that boiling water swells in volume. But I never connected my kitchen experience to the water level control in a kettle steam generator.

The boiler at the Valero Refinery in Delaware City produced erratically wet steam. The moisture in the steam caused the catalyst to partially plug the steam distributor in the fluid catalytic cracker spent catalyst stripper. The poor catalyst stripping resulted in reduced unit efficiency.

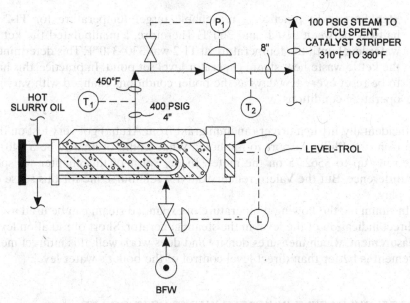

Figure 20-4 *Level control in a kettle waste heat steam generator*

I solved the wet steam problem by use of the Mollier diagram:

- **Step One**—The temperature of 415 psia saturated steam is 450 °F.
- **Step Two**—I note that the velocity at TI-1 is about the same as TI-2 (see Fig. 20-4). As long as velocity changes are less than 30–40 feet per second in steam lines, kinetic energy effects can be ignored.
- **Step Three**—Ignoring kinetic energy changes and ambient heat losses, the reduction in pressure across the P_1 control valve is an isoenthalpic expansion. Expanding the steam to 115 psia should result in 360 °F super-heated steam.
- **Step Four**—The observed temperature at TI-2 is not 360 °F, but 320 °F because of entrained moisture in the 400 psig steam.
- **Step Five**—The amount of water in the 400 psig generated steam is calculated:

$$(360\,°F - 320\,°F) \cdot (0.55) \div 980 = 2.3\,wt.\%$$

(The 940 BTU/lb is the latent heat of steam at 100 psig.)

- **Step Six**—The 2.3 wt.% water in the 400 psig steam is still okay because it will all evaporate in the 100 psig steam. (In this case I'm not concerned about salts.) The saturation temperature of 100 psig is 310 °F. So, once the temperature of the 100 psig steam falls to 310 °F, the steam supply to the catalyst stripper will be wet.

- **Step Seven**—In practice, a reasonable target temperature for TI-2 is halfway between 310 °F and 360 °F. Therefore, I manipulated the kettle water level until the temperature at TI-2 was 330–340 °F. This determined the kettle waste heat steam generator level set point. In practice, this had to be reset every few days as the boiler conditions changed with varying operation conditions.

Incidentally, my taxi driver (an immigrant from Aruba) got very lost on the trip from the Philly Airport to my hotel. What would ordinarily be a $40.00 fare rang up to $562.75 on the meter. In the spirit of this chapter we split the difference. But the Valero refinery did not reimburse me for this sense of fair play.

In summary, the flowing temperature of expanded steam can be used as an indirect indication of the level in the steam generator. Short of radiation level measurement, which measures density and does work well, this indirect measurement is better than direct level control of the boiler's water level.

OVERFLOW BAFFLE IN KETTLE WASTE HEAT BOILER

Steam boilers that drive process pumps have been in use for 300 years. Thomas Newcombe used steam to drive his reciprocating beam pump in 1720 to drain flooded coal mines in England. One would think that nothing new could be written about control of boilers generating medium-pressure steam. One would think that all control problems associated with the design and operation of an ordinary waste heat boiler would have been solved many generations ago.

Not true! Many plant operators suspect that there are unresolved problems with boiler control. I'm referring to the control of the water level inside the boiler. Applying the level control principles that I will explain will improve the operating integrity and efficiency of most process plants. Also, it's an example of how to use the principles of thermodynamics to solve practical process control problems.

Overflow Baffle in Kettle Boiler

Perhaps the most common method to deal with level control on a waste heat kettle-type steam generator is shown in Figure 20-5. This boiler relies on a baffle to maintain a proper level in the kettle. Sometimes the water inlet is on either side of the baffle. Sometimes the level indication is also on either side of the baffle. I have maintained for 44 years that this baffle serves no purpose. No one I have ever met has understood its purpose, either. Most plant maintenance divisions have cut a hole in the bottom of the baffle. This hole defeats any potential purpose of having such a baffle. Despite the widespread and

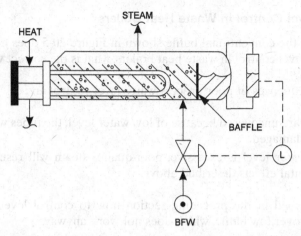

Figure 20-5 *Baffle in kettle waste heat boilers are design errors and serve no purpose*

almost universal use of this baffle feature in boilers, it represents an engineering control design error. I'll explain with reference to Figure 20-5.

First note that there is no relationship between the water level to the right of the baffle and the boiling water level to the left side of the baffle until the water reaches the top edge of the baffle. Only at this elevation can we hope to correlate the two fluid levels. However, once the water level reaches the top edge of the baffle, we could remove the baffle without affecting the level control to the left side of the baffle.

Some installations have the water entering to the right side of the baffle. But then the water must overflow the baffle before it influences the level where steam generation takes place. If the water must overflow the baffle before it can affect the level on the left of the baffle, then the baffle is always submerged and cannot influence level control.

If the level control is located to the left side of the baffle and water is introduced to the right of the baffle, then certainly the baffle is not influencing the level indication in the boiler.

The reason for this lengthy treatment of the useless baffle is threefold:

- Most waste heat kettle boilers have this totally useless feature.
- In my seminars I am constantly asked how this baffle works, and why the baffles are always retrofitted with holes cut into the bottom.
- It's a nice example of how the failure to grasp fundamental process control concepts results in self-perpetuating and expensive process control errors. By expensive I refer to the added length of the shell to accommodate the baffle and end chamber section.

Correct Level Control in Waste Heat Boilers

If the use of the conventional baffle shown in Figure 20-5 does not solve the problem of level control in waste heat boilers, what is the correct way to solve this problem?

The consequences of poor process water level control are:

- If tubes are uncovered because of low water level, the tubes will overheat and be damaged.
- If the water level is excessive, poor-quality steam will result, with the detrimental effects described above.

I've described in the preceding section how to control levels in boilers without this overflow baffle, which does not work anyway.

LEVEL CONTROL IN DEAERATORS

Ordinary level control in a boiler feed water deaerator is simple and conventional—up to a point! This point is when the steam pressure control valve opens to 100%. To understand the relationship between pressure control and level control in a deaerator I should first explain how a deaerator works.

The purpose of a deaerator is to steam strip air out of boiler feed water (BFW). Oxygen is extremely corrosive to boiler tubes and thus must be removed from the BFW. The top portion of the deaerator shown in Figure 20-6 is a small-trayed steam stripper tower. The lower portion is the reservoir portion. Steam mixes with the cold BFW in the small stripping section. Before the steam can strip out any air, the cold water must be heated to the boiling point temperature of water at the deaerators' operating pressure. Some steam always escapes through the top vent, which is a fixed orifice. The rest of the steam is consumed in maintaining the BFW at its boiling point temperature and pressure.

If there is insufficient steam flow to accomplish this objective, the deaerator pressure and temperature will fall. Note that the water in the lower portion of the vessel will always be saturated water at its boiling or bubble point. As the amount of cold water entering the deaerator needed to maintain the deaerator level increases, more steam is needed to heat the cold water to maintain deaerator pressure. But what happens when the steam pressure control valve opens to 100%? From a bad experience, I can tell you exactly what happens:

- The pressure in the deaerator will drop sharply.
- The water in the reservoir section will begin to boil.
- As the water boils, its density rapidly falls.

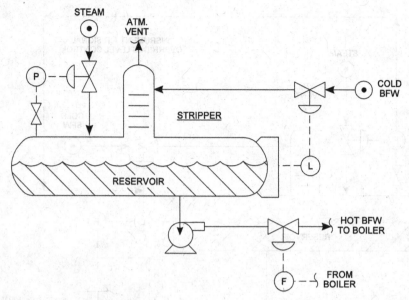

Figure 20-6 Level control problem in deaerator. Cascade control of pressure to level is needed for safety

- The deaerator level indicator interprets this drop in fluid density as a reduction in water level.
- The cold water makeup flow increases to restore the water level in the reservoir, even though the water level is not actually low. If anything, the boiling water level is rapidly rising.
- The additional flow of cold BFW drives down the deaerator temperature and pressure. The PRC (pressure control valve) on the steam supply cannot open any further because it's already 100% open.
- The falling reservoir pressure reduces the density of the boiling water. The level indicator interprets this drop in density as a further reduction in level and calls for more cold BFW, which further drives down the deaerator pressure and temperature.

This positive feedback control loop will continue to feed upon itself until the water explodes out of the top vent. Or the deaerator pressure may drop below atmospheric pressure. The resulting partial vacuum can and has caused vessels to collapse.

Deaerator Safety Override

Certainly we would like to prevent this hazardous condition from developing. This is done by cascade override control (see Chapter 12). Once the pressure

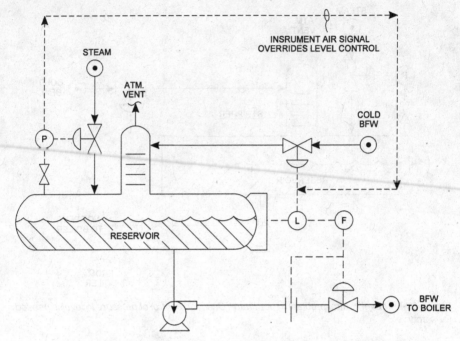

Figure 20-7 *Override pressure control on a boiler feed water deaerator*

control valve shown in Figure 20-6 is 100% open, it must override the cold BFW makeup level control valve. The objective is to prevent the cold water makeup BFW control valve opening any further. Of course, the deaerator level will start to fall, and an alarm should sound. The boiler plant operators must now take the appropriate action at their boilers to reduce the hot BFW demand from the deaerator reservoir. However, once the deaerator level falls below a certain point, the level control must cascade and override the flow control valve to protect the hot BFW pump from cavitation due to loss of suction pressure. Cavitation will damage the mechanical seals of the pump because of vibration and the dual seal faces overheating. Figure 20-7 shows how the pressure control valve on the steam supply will override the normal level control valve on the cold BFW water supply to the deaerator. An additional override loop (not shown) will be needed to protect the pump from cavitation.

Control of Boiler Blowdown

The purpose of maintaining a continuous flow of purge water from the bottom of a boiler is to purge the boiler of salts. The concentration of these salts in the boiling water is called TDS (total dissolved solids). The maximum allowable concentration of TDS is a function of the steam pressure generated.

Higher-pressure steam requires a lower TDS in the boiler, because the tubes run hotter and can be more easily damaged by salt deposit.

The quantity of blowdown water is mainly a function of the quality of the BFW. BFW produced in a modern, well-operated demineralization plant likely requires several percent blowdown. Percent blowdown is the ratio of the amount of water drained from the boiler to the BFW rate flowing to the boiler.

If the BFW comes from an old, hot line softening plant, at least 10% of blowdown likely will be required to keep salts in steam acceptable.

A TDS analyzer can be used to continuously and automatically control the blowdown rate. Excessive blowdown water rates waste energy and swell the plant's water effluent flow. Too little blowdown scales the boiler tubes and promotes poor-quality steam and carryover of hardness deposits to downstream steam turbines and fired heater steam superheater tubes (see Chapter 8, "Analyzer Process Control").

21

Level, Pressure, Flow, and Temperature Indication Methods

There is an implied assumption in this text. That is, I have assumed that the reader is familiar with how the process equipment that is to be controlled actually works. For those readers that are not familiar with the function of:

- Heat exchangers
- Fired heaters
- Compressors
- Turbines
- Knock-out drums
- Distillation towers
- Air coolers
- Centrifugal pumps
- Vacuum system

I have written a book in conjunction with E. T. Lieberman, published by McGraw Hill, *A Working Guide to Process Equipment—3rd Edition*[1]. Sections of this chapter have been adapted from this prior publication. For me, the understanding of the control of process equipment and the understanding of how such equipment functions are just two aspects of the same subject.

Troubleshooting Process Plant Control, by Norman P. Lieberman
Copyright © 2009 John Wiley & Sons, Inc.

What is the difference between a gauge glass and a level glass? Simple! There is no such thing as a level glass. The liquid level shown in a gauge glass does not correspond to the level in a process vessel. Figure 21-1 is a good example. This is the bottom of an amine fuel gas absorber. This tower is used to remove hydrogen sulfide from the fuel gas. At the bottom of the tower there are three phases:

- Fuel gas: 0.01 specific gravity
- Hydrocarbon liquid: 0.60 specific gravity
- Rich amine: 0.98 specific gravity

Because of the location of the level taps of the gauge glass, only the amine is in the glass. The gauge glass simply measures the pressure difference between two points of the tower (points A and B in Fig. 21-1). That is, the gauge glass functions as a monometer that measures the pressure difference in terms of the specific gravity of the liquid in the gauge glass. Should the specific gravity of the liquid in the glass be the same as that of the liquid in the tower, both the gauge glass level and the tower level would be the same. But this is never so. The specific gravity of the liquid in the gauge glass is always greater than the specific gravity of the liquid in the tower. Hence, the apparent liquid level in the gauge glass is always somewhat lower than the actual liquid level in the tower.

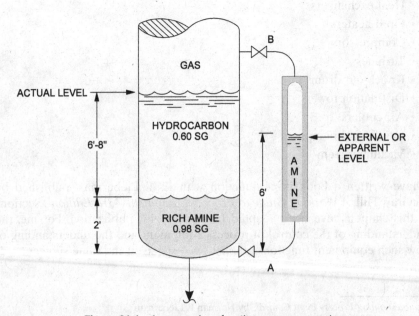

Figure 21-1 A gauge glass functions as a manometer

This discrepancy between the apparent level in the gauge glass and the actual level in the tower also occurs in any other type of level-measuring device. This includes external float chambers, displacement chambers, and level-trols. The one exception to this is level-measuring devices using radiation techniques.

The three causes of the discrepancy between the external level and the internal level are:

- Foam formation inside tower
- Ambient heat loss from the external gauge glass or level-trol
- The liquid specific gravity in the glass being greater than the specific gravity in the tower, as shown in Figure 21-1

Let's assume that the gauge glass shown in Figure 21-1 holds 6 feet of amine. Since the bottom tap is in the amine phase and the top tap is in the gas phase, the liquid hydrocarbon is excluded from the gauge glass. To balance out the weight of the 6 feet of amine, the tower would have to have about 2 feet of amine and 6 feet 8 inches of liquid hydrocarbon. That is, the tower liquid level would be about 8 feet 8 inches or 2 feet 8 inches higher than the gauge glass level.

If you conclude from the above that we could use the gauge glass level to actually calculate the level inside the tower, you are quite wrong. To perform this calculation, one would have to assume the ratio of the phases. But this is an assumption equivalent to assuming the answer. How then does one determine the actual liquid level in the tower on the basis of the apparent liquid level in the gauge glass? The answer is that there is no answer. It cannot be done! And this statement applies to all other sorts of level-measuring instruments—with the exception of radiation devices.

EFFECTS OF TEMPERATURE ON LEVEL

The gauge glass will normally be somewhat colder than the process vessel as a result of ambient heat losses (an exception to this would be a refrigerated process). For every 100 °F decrease in the gauge glass temperature or level-trol temperature the specific gravity of the liquid in the glass increases by 5%. This rule of thumb is typical for hydrocarbons only. Aqueous (water-based) fluids are totally different.

For example, suppose the height of liquid in a gauge glass is 4 feet between the level taps. The glass temperature is 60 °F. The tower temperature is 560 °F. How much higher is the height of liquid in the tower than the glass? I have calculated the solution:

$$\frac{500\,°F}{100\,°F} \times 5\% = 25\%$$

- This means that the liquid in the gauge glass is 25% more dense than the liquid in the tower bottom.
- Assuming a linear relationship between density and volume, the level of liquid in the tower above the bottom tap of the gauge glass must be:

$$4 + 25\% \times 4\,ft = 5\,ft$$

- In other words, the liquid in the tower is 1 ft above the level shown in the glass.

PLUGGED TAPS

How do plugged level-sensing taps affect the apparent liquid level in a vessel? Let's assume that the vapor in the vessel could be fully condensed at the temperature in the gauge glass. If the bottom tap is closed, the level will go up because the condensing vapors cannot drain out of the glass. If the top tap is closed, the level will go up because the condensing vapors create an area of low pressure, which draws the liquid up the glass through the bottom tap. Thus, if either the top or bottom taps plug, the result is a false high-level indication (see Chapter 22, "Alarm and Trip Design for Safe Plant Operations").

HIGH LIQUID LEVEL

In our calculation above, we had 4 feet of liquid in the glass and 5 feet of liquid in the tower. But what happens if the distance between the two taps is 4 feet 6 inches? I have drawn a picture of the observed result in Figure 21-2. Liquid circulates through the glass, pouring through the top tap and draining through the bottom tap. The apparent liquid level would then be somewhere between 4 feet 0 inches and 4 feet 6 inches, let's say 4 feet 2 inches. The indicated liquid level on the control room chart would then be 92% (i.e., 4 feet 2 inches ÷ 4 feet 6 inches) As the liquid level in the tower increases from 5 feet to 100 feet, the indicated liquid level would remain at 92%.

Once the actual liquid level inside the tower bottom rises above the top-level tap, no further increase in level can be observed in the gauge glass. We say the level indication is "tapped out."

The same sort of problem arises in a level-trol, which measures and transmits a process vessel liquid level to the control center. As shown in Figure 21-3, the level-trol operates by means of two pressure transducers, devices for converting a pressure signal into a small electric current. The different between the two pressure transducers shown in Figure 21-3 is called the milliamp output. Output is proportional to the pressure difference between the bottom and top taps in the level-trol. To convert the milliamp output signal from the

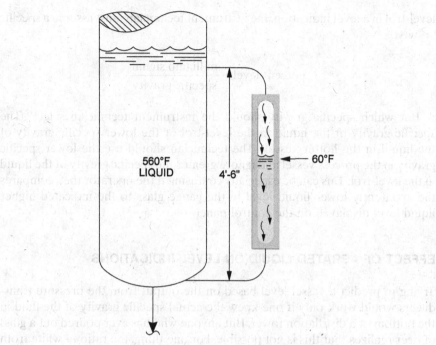

Figure 21-2   *The circulation of liquid in a gauge glass*

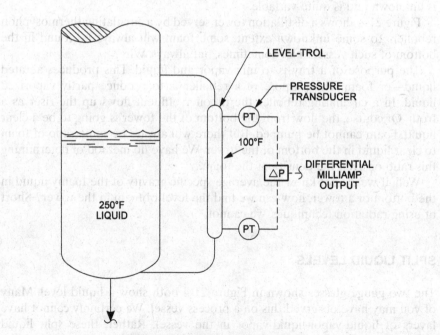

Figure 21-3   *Operation of a level-trol*

level-trol in a level indication, the instrument technician must assume a specific gravity:

$$\text{percent level} \sim \frac{\text{milliamp signal}}{\text{specific gravity}}$$

But which specific gravity should the instrument technician select? The specific gravity of the liquid in the level-trol or the lower specific gravity of the liquid in the hotter vessel? The technician should use the lower specific gravity in the process vessel and ignore greater the specific gravity of the liquid in the level-trol. This can be especially confusing if the operator then compares the apparently lower liquid level in the gauge glass to the indicated higher liquid level displayed on the control panel.

EFFECT OF AERATED LIQUID ON LEVEL INDICATIONS

Trying to predict a vessel level based on the output from the pressure transducers would work only if one knew the actual specific gravity of the fluid in the bottom of a distillation tower. But anyone who has ever poured out a glass of beer realizes that this is not possible. For one thing, the ratio of white froth to yellow beer is never known in advance. Also, the density of the froth itself is unknown and is quite variable.

Figure 21-4 shows a distillation tower served by a circulating thermosyphon reboiler. To some unknown extent, some foam will always be found in the bottom of such vessels. Not sometimes, but always. Why?

The purpose of a tray is to mix vapor and liquid. This produces aerated liquid—or foam. The purpose of a reboiler is to produce partly vaporized liquid. In a circulating reboiler the reboiler effluent flows up the riser as a froth. Of course, the flow from the bottom of the tower is going to be a clear liquid. Foam cannot be pumped. But there will always be some ratio of foam to clear liquid in the bottom of the tower. We have no method of determining this ratio or even the density of the foam.

Well, if we do not know the average specific gravity of the foamy liquid in the bottom of a tower, how can we find the level of foam in the tower? Short of using radiation techniques, we cannot.

SPLIT LIQUID LEVELS

The two gauge glasses shown in Figure 21-4 both show a liquid level. Many of you may have observed this on a process vessel. We certainly cannot have layers of liquid-vapor-liquid-vapor in the vessel. Rather, these split liquid levels are a positive indication of foam or froth in the bottom of the tower.

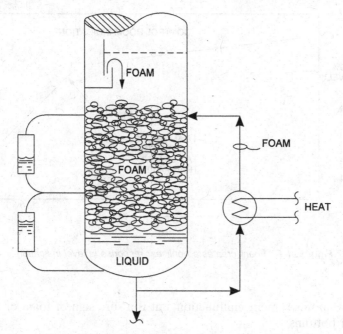

Figure 21-4   *Split liquid level indication caused by foam*

If the foam is spanning both taps of a gauge glass, then the height of the liquid in the glass is a measure of the specific gravity or density of the foam in terms of the specific gravity of the liquid in the glass. If the foam is above the top tap of both the gauge glasses then there will be a level in both glasses. The upper gauge glass will show a lower level because the light foam in the tower floats on the top of the heavier foam. Note that these split liquid levels, so often seen in a process vessel, tell us nothing about the real liquid level in the vessel. They are a sign of foam.

Figure 21-5 is a plot of the liquid level in a crude preflash drum versus time. We were steadily withdrawing 10% more flashed crude from the bottoms pump than the inlet crude feed rate. The rate of decline of the liquid level noted in the control center was only about 25% of our calculated rate. Suddenly, when the apparent level in the control room had reached 40%, the level indication started to decline much more rapidly. Why?

This extreme nonlinear response of a level to a step change in a flow rate is quite common (see Chapter 23, "Nonlinear Process Responses"). Before the sudden decline in the indicated liquid level, foam had filled the drum above the top-level tap. The initial slow decline in the apparent level was due to the denser foam dropping between the level taps being replaced by a lighter foam. Only when the foam level actually dropped below the top tap of the drum did the indicated liquid level begin to decline at a rate representing the actual decline in the level. Thus we can see that this common, nonlinear response is

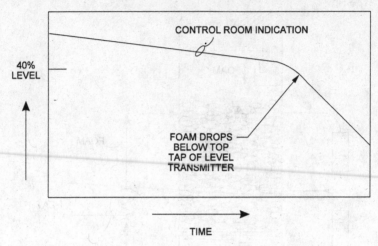

Figure 21-5 Foam creates a nonlinear response in level indication

not due to an instrument malfunction, but is a sure sign of foam or froth in the tower bottoms.

RADIATION LEVEL DETECTION

The only way around the sort of problems discussed above is to use neutrons or X-rays to measure the density in a vessel. The neutron backscatter technique is best performed with hydrogen-containing products. Both the source of slow neutrons and the receiver are located in the same box. The slow neutrons bounce off of protons (hydrogen ions) and are reflected back. The rate at which these neutrons are reflected back is measured and corresponds to the hydrocarbon density in the vessel. This measurement is not affected by steel components inside—or outside—the vessel.

X-ray level detection works with a source of radiation and a receiver, located on either side of the vessel. As the percent absorption of the radiation increases, the receiver sees fewer X-rays and a higher density is implied. The X-rays are absorbed by steel components such as ladders and manways, which can sometimes be confusing.

Either method discriminates nicely between clear liquid, foam, and vapor. Such a level controller can be calibrated to hold a foam level or a liquid level. Of course, this sort of radiation level detection is far more expensive than conventional techniques.

VACUUM PRESSURE MEASUREMENTS

A West Texas process plant had decided to replace the main condenser. Colder weather always coincided with a vastly improved vacuum in their vacuum tower. It seemed as though colder air to the condenser really helped. So the control engineer concluded that a bigger condenser would also help during warm weather.

Wrong! The engineer failed to realize that the vacuum pressure indicator was not equipped with a barometric pressure compensator. An ordinary vacuum pressure indicator or pressure gauge reads the pressure difference between the vacuum system and atmospheric pressure. When ambient temperatures drop, the barometer rises or ambient pressure goes up. An ordinary vacuum pressure gauge or indicator would then read an improved vacuum. But in reality, the vacuum has not changed.

The opposite problem would occur in Denver—the Mile-High City. At sea level, full vacuum is 30 inches of mercury (or 30 inches Hg), but in Denver full vacuum is about 26 inches Hg. An ordinary vacuum pressure gauge reads 0 inches Hg in Denver and in New Orleans, because although these cities are at different altitudes the vacuum pressure gauge compares system pressure only with ambient pressure. But a vacuum pressure gauge reading 25 inches Hg in New Orleans would correspond to a poor vacuum of 5 inches Hg absolute pressure (30 inches Hg – 25 inches Hg). A vacuum pressure gauge reading 25 inches Hg in Denver would correspond to an excellent vacuum of 1 inch Hg absolute pressure (26 inches Hg – 25 inches Hg).

All these complications can be avoided when making field measurements by using the vacuum manometer shown in Figure 21-6. The difference between

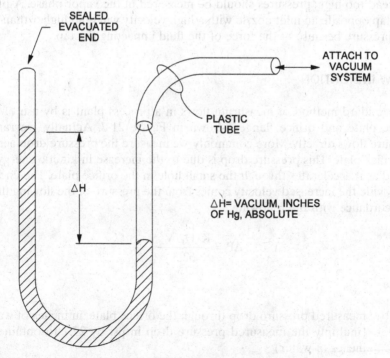

Figure 21-6 A mercury absolute-pressure manometer

the two mercury levels is the correct inches of mercury absolute pressure or millimeters of mercury (mmHg). Be careful! A drop of water on the evacuated end of the manometer will result in a falsely low vacuum reading, because of the vapor pressure of water.

Pressure Transducers

Disassemble a pressure transducer and you will see a small plastic diaphragm. A change in pressure distorts this diaphragm and generates a small electrical signal. The signal must be quite tiny, because placing your hand on the transducer can alter its reading. A modern digital pressure gauge uses a pressure transducer. This type of gauge, if zeroed at sea level in New Orleans, Louisiana, will read 4 inches Hg vacuum in Denver, Colorado. Most pressure signals transmitted from the field into the control center are generated from pressure transducers. Differential pressure indicators simply take the differential readings from two transducers and generate a milliamp output signal.

Location of Pressure Tap

Locating a pressure tap in an area of high velocity is likely to produce a lower pressure indication than the real flowing pressure. Using a purge gas to keep a pressure tap from plugging often can cause a high pressure reading if too much purge gas or steam is used. A pressure tap located below a liquid level will read too high; pressures should be measured in the vapor phase. A pressure tap opposite an inlet nozzle with a high velocity will read higher than the real pressure because of the force of the fluid impacting the tap.

FLOW INDICATION

The standard method of measuring flows in a process plant is by use of the orifice plate and orifice flanges, shown in Figure 21-7. Actually, we rarely measure flows directly. More commonly, we measure the pressure drop across an orifice plate. This pressure drop is due to the increase in kinetic energy of a fluid as it accelerates through the small hole in the orifice plate. The energy to provide the increased velocity comes from the pressure of the flowing fluid, in accordance with the following:

$$\Delta P = \frac{K \cdot D_f \cdot V^2}{62.3}$$

where

ΔP = measured pressure drop through the orifice plate, in inches of water (multiply the measured pressure drop in psi by 27.7 to obtain the inches of water)

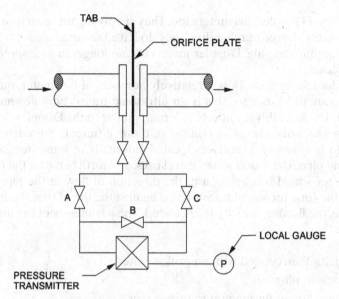

Figure 21-7 *Orifice flowmeter*

V = velocity of the fluid through the orifice plate, ft/s

D_f = density of the fluid, whether vapor or liquid, lb/ft^3

K = an orifice coefficient

You should look up the orifice coefficient K in your Cameron or your Crane[2] handbook—but it is typically 0.6.

NON-ORIFICE-TYPE FLOW MEASUREMENT METHODS

There are several other methods in common use to measure process flows. Most of these methods depend on measuring differential pressure drop across a restriction. For example, the wedge meter is used in severe plugging services such as delayed coker furnace feed. It is rugged and reliable unless the orifice taps plug. The wedge meter itself is no more prone to plugged pressure taps than an orifice plate-type flow meter. It's just that the services that the wedge meters are used in are themselves prone to plugging the pressure tap connections.

Venturi meters are often used in the suction of large air blowers. They are reasonably accurate, but their main advantage is that they develop very little pressure drop (0–10 inches of water). A conventional orifice-type flow indicator will typically develop a delta P of 0–100 inches of water. Venturi meters also rely on dual-pressure taps, which are subject to plugging.

I have tried Doppler flowmeters too. They simply do not seem to work with any reasonable degree of reliability. They do have the advantage of not having any pressure taps to plug. Doppler meters are no longer in widespread use in process plants.

What does seem to work is a relatively new sort of flowmeter, introduced by GE about 10 years ago. This is an ultrasonic transit time flowmeter (see Fig. 21-8). The assembly is entirely external, similar to the Doppler flowmeter. However, the principle of operation is quite different. Sound waves are exchanged between two receivers located externally at some distance along the flowing pipe. The sound waves travel back and forth between the receivers. The time for sound transmission in the direction of flow in the pipe is compared to the time for sound transmission against the flow. From the measured delta time, the flowing velocity is calculated. The advantages of the ultrasound meter are:

- No delta P associated with an orifice plate
- No taps to plug
- No need to use flushing oil to orifice taps
- Nothing to freeze up

This relatively new and novel technology is gradually replacing wedge meters in difficult refinery services, such as delayed cokers and visbreakers. For small flows, I've used a magnetic rotometer. A steel magnetic ball is lifted inside a nonmagnetic tube by up-flowing fluid velocity. A magnetic ring outside the tube follows the ball's position. I've used this simple and very rugged device to measure sulfuric acid and caustic flows with great confidence.

I've also used pilot tubes as a trend indicator. These are simple tubes inserted at right angles to a flow. They will give an approximate idea of a flow in a large-diameter line with very little cost or pressure loss. They are used in process plant flare lines to measure occasional spills of gas to the plant flare system. The holes in the pilot tube itself are, however, subject to plugging.

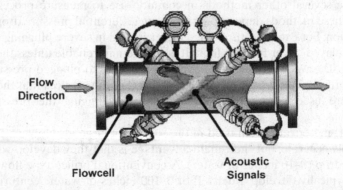

Figure 21-8 Ultrasonic transit time flow measurement technology (courtesy GE Sensing)

CHECKING FLOWS IN THE FIELD

The competent engineer does not assume a flow indication shown on the control panel is correct but proceeds as follows:

1. Referring to Figure 21-7, place a pressure gauge in the position shown. I like to use a digital gauge.
2. By opening both valves A and B, with C closed, you will now be reading the upstream pressure.
3. By opening valve C, with A and B closed, you will read the pressure downstream of the orifice plate.
4. The difference between the two readings is ΔP in the preceding equation. Now solve for V in the equation.
5. Look at the tab sticking out of the orifice flanges (see Fig. 21-7). If the orifice plate is installed in the correct direction, there will be a number stamped onto the tab, facing toward the flow. This is the orifice plate hole diameter. For example, if you see 0.374″ stamped on the tab, the orifice hole diameter should be 0.374 inches.
6. Using the hole diameter and V, calculate the volume of fluid

The instrument mechanic has filled the lines with glycol, mainly for winter freeze protection. Many process streams contain water, which can settle out at low points and, in effect, plug the impulse lines to flow- or level-sensing ΔP transmitters when water freezes. Note that there is not a lot of difference between measuring a flow and a level; they both are typically measured by using a differential pressure signal.

Naturally, just like level indicators, the flow orifice taps can plug. If the upstream tap plugs, the flow will read low or zero. It is best to blow the tap back with glycol, but that is not always practical. If you blow the taps out with the pressure of the process stream, you do not need to refill the impulse lines with glycol to get a correct flow reading. But the lines must be totally refilled with the same fluid. If you are measuring the flow of a single-phase liquid, just open valves A, B, or C (shown in Fig. 21-7) for a few minutes. If you are working with vapor at its dew point or wet gas, there is a problem. If the flow transmitter is located below the orifice flanges, you will have to wait until the impulse lines refill with liquid. Open valve B, and close valves A and C. Now wait until the flowmeter indication stops changing. It ought to go back to zero, if the lines are refilled.

CORRECTING FLOWMETER READING OFF-ZERO

The indicated flow of acetic acid is 9000 liters per day. The instrument technician checks the flowmeter to see whether it has drifted by opening valve B with A and C closed (Figure 21-7). It should go back to zero—but a reading of 2000 liters per day is noted. The full range on the flowmeter is 10,000 liters per day. What is the real flow rate of the acetic acid? The answer is not 7000 liters. Why? Because flow varies with the square root of the orifice plate pressure drop. To calculate the correct acetic acid flow:

$$9000^2 - 2000^2 = 77,000,000$$

$$(77,000,000)^{1/2} = 8780 \text{ liters per day}$$

The lesson is that near the top end of its range the indicated flow is likely to be accurate, even if the meter is not well zeroed or the measured delta P is not too accurate. On the other hand, flowmeters using orifice plates cannot be very accurate at the low end of their range, regardless of how carefully we have zeroed them. Digitally displayed flows also follow this rule.

You may notice when you measure delta P that if it is a small value, it is quite difficult to measure accurately. This means that the orifice plate hole is oversized, and that the accuracy of the recorded flow on the control panel is also poor. Or, if the measured delta P is quite high, then a lot of pressure is being wasted, and the orifice plate hole is undersized and restricting the flow. Furthermore, the recorded flow on the control panel may be off the scale.

The reason the orifice flanges are kept close to the orifice plate is that when the liquid velocity decreases downstream of the orifice plate the pressure of the liquid goes partly back up. This is called pressure recovery. Whenever the velocity of a flowing fluid (vapor or liquid) decreases, its pressure goes partly back up. An extreme example of this is water hammer. The reason the pressure at the end of the pipe is lower than at the inlet to the pipe is frictional losses.

The orifice coefficient K takes into account both frictional pressure losses and conversion of pressure to velocity. The frictional losses represent an irreversible process. The conversion of pressure to velocity represents a reversible process.

TEMPERATURE MEASUREMENT

A thermocouple assembly is a junction consisting of two wires of different metallurgy. When this junction is heated, an electric current, proportional to the junction temperature, is produced. Different metal wires make up the three most common junctions: J, H, and K. It is not uncommon for a thermocouple, regardless of the type of junction, to generate too low a temperature signal.

If the exterior of the thermowell becomes fouled, the indicated temperature generated by the thermocouple will drop. The problem is that the external cap of the thermowell assembly radiates a small amount of heat to the atmosphere.

Normally this has a negligible effect on the indicated temperature. However, when the process temperature is 600 °F–800 °F, the thermowell is in a vapor phase and it becomes coated with coke; I have seen the indicated temperature drop by 40 °F below its true value. To verify that fouling of a thermowell is a problem, place a piece of loose insulation over the exterior thermowell assembly. If the indicated temperature rises by 50 °F or 10 °F, then fouling on the outside of the thermowell is proved.

For a thermocouple to read correctly, it should be fully inserted in a thermowell and the thermowell itself should extend several inches into the process liquid. If the process stream is a vapor, which has poorer heat-transfer properties than liquids, the thermowell, especially if the external vessel insulation is poor, should extend more than 6 inches into the process flow. To check the length of the thermowell, unscrew the thermocouple assembly and pull it out, then simply measure the length of the thermocouple. This is also a good opportunity to verify the control room reading with a portable temperature probe or a glass thermometer inserted in the thermowell. In general, the temperature indication displayed on the console is the most reliable of all the process variables. Most often, if they are reading different from an expected value, it usually indicates a localized process malfunction and not an instrumentation problem.

REFERENCES

1 Lieberman, E.T. & Lieberman, N.P., "A Working Guide to Process Equipment—3rd Edition". McGraw Hill, New York, 2008.
2 Crane Company "Flow of Fluids Through Valves, Fittings and Pipes," Technical Paper Number 410.

22

Alarm and Trip Design for Safe Plant Operations

Alarm points and trip parameters are two of the most important design components of any process plant. When any control variable moves beyond a predetermined range, normal automatic control is no longer sufficient. First, an alarm must alert the panel operator that the variable has moved beyond the acceptable control range. Then, the process must be automatically shut down by an independent trip when the required panel board operator intervention is not forthcoming.

The Process Control Engineer has a threefold function relating to alarms and trips. First, he must specify how the parameter to be alarmed will be measured. Second, the Process Control Engineer must decide what is an unacceptable value for any parameter. Finally, a test procedure must be specified to routinely prove that the alarm or trip is functional. A safety device that is not tested on a routine basis will never function in an emergency. You can imagine how I've become so smart on this subject.

THE CONCEPT OF REDUNDANCY

The excess pressure safety relief valve had just opened. The refinery flare was blazing away. My alkylation unit depropanizer was releasing vast quantities of propane vapor through its dual 8-inch relief valves. This had all happened quite suddenly, with no warning whatsoever. Why hadn't the high pressure alarm

---

sounded to alert the panel operator that the depropanizer tower was exceeding its maximum permissible range of operating pressure?

Figure 22-1 illustrates the problem. I had been injecting a corrosion inhibitor chemical purchased from Petrolite into the tower's reflux. The chemical was dissolved in water, which evaporated in my depropanizer. The residue was black and sticky. With time the tower instrument taps, including the pressure-sensing tap located below the bottom tray, began to plug. When this ¾-inch connection plugged off entirely, the propane pressure to the PRC began to drop. Referring to Figure 22-1, note that there is a line drawn under the letters PRC. This line indicates that the instrument output is displayed on the panel. Without this line under the PRC the pressure indication would only be displayed locally in the field. The letters themselves stand for:

- P = Pressure
- R = Recorded on the panel
- C = Controller (i.e., controlling the steam flow to the reboiler)

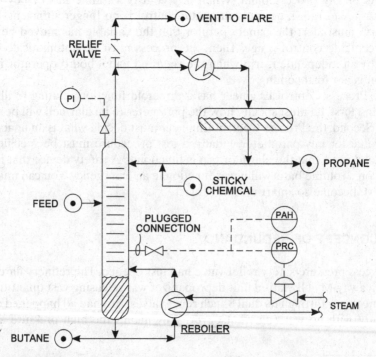

Figure 22-1 *High pressure alarm senses lower pressure from same connection as the pressure controller. This is bad design practice*

The PAH instrument is also displaced on the panel. The letters stand for:

- P = pressure
- A = alarm displayed on the panel
- H = high pressure

As the measured tower pressure dropped, because of the black sticky liquid plugging the pressure connection, the steam flow to the reboiler increased. The extra steam increased the tower pressure, but not the pressure of the plugged ¾-inch pressure connection. The high pressure alarm failed to sound because it was not connected to a separate pressure tap connection.

This simple story indicates the need for redundancy in measuring a process variable such as pressure. The instrument failure that caused the control malfunction was the same failure that prevented the alarm from sounding. The lesson for the process control student is the need for redundancy. In this case, the control sensing point should have been completely separated from the alarm sensing point. Had there been a safety trip involved (i.e., to automatically shut off the heat to the reboiler upon sensing the excessive pressure), then we should require a third sensing connection to activate the high depropanizer pressure trip valve mechanism.

After this flaring incident, which generated a black plume over Texas City, I relocated the pressure alarm sensing point. Figure 22-1 shows a PI point above the top tray. There is no line under the letters, so it's a local pressure indication. The letter "I" stands for indication only, without a recorder, that is, just an ordinary pressure gauge. I had the pressure alarm connected to the same connections as this pressure gauge tap. I kept the pressure gauge at this location, too by using a tee junction. Thus I had provided true redundancy to this pressure sensing system.

TESTING THE NEW PRESSURE ALARM

When I pass on and the Creator asks me what I have learned during my life on Earth, I'll say, "Master of the Universe, a safety device that is not routinely tested will never function in an emergency." I've had to learn this lesson a dozen times, but I have learned it very well. Testing an instrument safety device is a three-step procedure. I'll illustrate these steps with reference to the relocated high pressure alarm on my alkylation unit depropanizer in Texas City:

- **Step One**—I tested the sensing point. That is, I noted that the local pressure gauge was reading 300 psig. I removed the gauge, and when I replaced it the pressure indication returned quickly to 300 psig. This proved that my pressure sensing point was not plugged. If the gauge needle requires more than 15 seconds to return to 300 psig, then the pressure guage connection is partly plugged.

- **Step Two**—Naturally, I did not want to actually raise the tower pressure above its maximum permissible operating range just to test the high pressure alarm function. Using a bottle of nitrogen, Mondo Lira, my instrument technician, applied 320 psig to the pressure alarm transmitter. The depropanizer alarm was calibrated to alarm at 20 psig below the relief valve set point.
- **Step Three**—I verified that the red light for the pressure alarm was lit on the control panel and that the audible high pressure alarm had sounded in the control center.

LEVEL ALARMS

Mondo Lira was a full-blooded Navajo Indian. In the hidden recesses of his mind was stored the wisdom of the ages. But to access this knowledge you had to ask the right questions. Mondo was absolutely silent, unless asked a specific question.

"Mondo, why did the boiler drum go dry, even though the low water level alarm didn't light on the control room alarm panel?" I asked. "Look, the boiler drum is still empty, but the low level alarm is still off."

"Because you people are stupid," Mondo concisely replied. All his advice was always prefaced by this observation of universal stupidity.

"Mondo, could you please be more specific? Stupidity is pretty rampant in this refinery."

"The level instrument setup on the boiler steam drum is stupid," he answered.

"But just how?" I asked.

"Mr. Norman Lieberman, Unit Operating Superintendent, do you want the whole story?" asked my Native American I and E tech.

"Yes, Mr. Mondo Lira, Senior East Plant Instrument and Electrical Technician. Yes, I wish to hear the whole story," I answered.

"Very well. I will speak and you will listen and learn," Mondo concluded in an authoritative tone.

Mondo drew two sketches for me: Figure 22-2 and Figure 22-3. The first sketch represented our current "stupid" design. Mondo explained, "Suppose connection B plugs on the level-trol. Steam will be drawn through the top tap of the level-trol and condense to water. The water level will rise in the level-trol and it can't drain back into the drum. Or suppose connection A plugs. The steam will condense in the level-trol and create a low pressure in the level-trol. Water will then be drawn out of the drum through connection B. Either plugged level connection will cause the level in the level-trol to rise above the water level in the drum. Either way the level control valve will close. Thus the actual water level in the drum will drop. Then the drum will go dry and empty. But the low level alarm will not be activated! Why? Because the low level alarm transmitter is connected to the same level connection taps as the level-

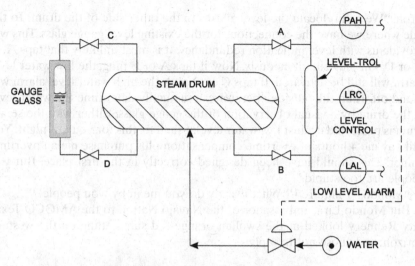

Figure 22-2   *Improper level instrumentation*

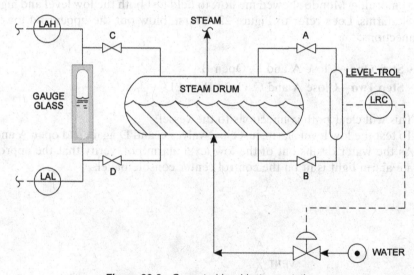

Figure 22-3   *Corrected level instrumentation*

trol. Hence, Superintendent Norman Lieberman, the low level alarm is subject to the same malfunction as the level-trol. Both transmitters will fail together at the same time and for the same reason. The problem is that you people are stupid. There is no redundancy."

"Okay, Mondo," I interjected. "Okay, but what shall we do to fix it?"

"Just so. I will enlighten you. I will set your feet on the path to wisdom and understanding. I will draw your eyes to Figure 22-3." Mondo was definitely on a roll. "We will relocate the level alarms to the other side of the drum. To the side where we have the connections for the existing level gauge glass. This will provide us with level indication redundancy. It is most unlikely that taps A, B, C, or D will plug simultaneously. Now, if taps A or B plug, the low water level alarm will still be activated. If taps C or D plug, the high water level alarm will sound off. Then the outside operator can manually determine the water level in the drum by a visual observation of the gauge glass. Either way, the steam drum is protected against low water level. I shall fix this for you this night. You will pay me 4 hours of overtime compensation plus purchase me an overtime dinner. This should have been designed correctly in the first place. But you people are too stupid."

"Mondo," I inquired. "What exactly do you mean by 'you people'?"

But Mondo Lira, ambassador of the Navajo Nation to the AMOCO Texas City Refinery looked at the swollen orange-red sun settling on the western horizon, and chose not to reply.

Testing Liquid Level Alarms

Next morning, Mondo showed me how to field test both the low level and high level alarms. Let's refer to Figure 22-4. First, blow out the upper and lower connections:

- **Step One**—Close A and D. Open B.
- **Step Two**—Close A and C. Open B and D.

This will clear both connections to the vessel.

To test the low level alarm, just close valves C and D again and open A and B. As the water drains out of the low level alarm pot, verify that the appropriate alarm light is lit on the control center console panel.

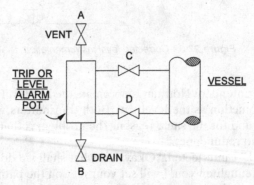

Figure 22-4 Level alarm or trip configuration

To test the high level alarm, connect a water hose to B. Valves A and B are open. Valves C and D are closed. As the water overflows the vent, verify that the appropriate alarm sounds in the control room.

Field Testing a Level Trip

Let us assume that we wish to protect the steam drum shown in Figure 22-3 from overfilling. This is accomplished by having a third level pot with dual taps connected directly to the drum. The output from the transmitter from this level pot will go to the water makeup level control valve. Should the level exceed a safe range, the level trip pot output will completely "trip off" (i.e., completely shut) the water flow to the steam drum.

This works with a mercuroid switch rather like the one you have in your thermostat at home. The mercury slides across a glass tube to open or close an electric circuit. The resulting current then activates a solenoid valve, which shuts off the water flow. This is done either with a separate shutoff valve or by closing the level control valve shown in Figure 22-3.

Of course, when testing the high-level trip for operability, we would not actually want to shut off the flow. We would therefore test the trip circuit as follows:

- **Step One**—Referring to Figure 22-4, attach a water hose to B.
- **Step Two**—Mechanically jam the shutoff valve. I would use a wooden wedge in Texas City. The obstruction must permit the shutoff valve to partly close from its initial condition.
- **Step Three**—Close C and D. Open A and B.
- **Step Four**—Fill the level pot with water from the hose.
- **Step Five**—Observe that the shutoff valve closes against the wooden wedge. One can safely assume that if the valve closes partway, it would then close all the way during an actual trip situation. That is, you have proved that the trip valve is not stuck.

USE OF CONDUCTIVITY PROBE

In a refinery we often have to be cautious about a slug of water entering a low-pressure distillation tower. Water will expand with explosive force and damage the tower internals, especially in larger-diameter columns. To protect the distillation tower trays from such a pressure surge, a conductivity probe may be inserted in the process flow. Hydrocarbon has a very low conductivity. The sudden increase in conductivity would shut a valve to stop the flow automatically. The trip circuit can be tested in a manner similar to the high-level trip test just described. The main difference is that the probe would be withdrawn through a packing gland and placed in a pail of water. A packing gland is a simple mechanism used to allow a thin probe (perhaps half-inch diameter) to be extracted from a process line flowing under pressure.

HIGH TEMPERATURE TRIPS

Usually a temperature indicator reads correctly, or reads so far from its expected value that the malfunction is obvious. But let's say we wish to verify the operational integrity of a high temperature trip. To be honest, I have not seen anyone test the operational integrity of any temperature trip circuit in 40 years. Yet I do know of very serious accidents that have happened when a high temperature trip failed to function.

- **Step One**—Heat an oil bath (for moderate temperature) or a pail of sand (for higher temperature) to the trip temperature with an electric heating coil.
- **Step Two**—Pull out the thermocouple assembly (not the thermowell) and insert it into the heated oil or sand. Incidentally, the thermowell should have been seal welded, not just screwed.
- **Step Three**—The rest of the procedure follows that just described in the preceding two sections.

Stanley, the old instrument tech on No. 12 Pipe Still at the Amoco Refinery in Whiting, Indiana, taught me a dozen such methods in the 1960s. Even then, Stan was a monument to arcane refinery practices. I've forgotten Stanley's last name, but I remember everything else that he taught me.

FLOWS

Typically, we may want a valve to shut off on low flow. An example would be tripping off furnace fuel gas upon low process feed flow. In addition, we might want the emergency steam flow to open automatically to purge out the furnace tubes upon loss of feed flow before the plant heater radiant tubes are coked off. Let's refer to Figure 22-5:

- **Step One**—Inform the console operator as to what is about to take place. I often forget this step. This advice also applies to all the preceding sections of this chapter.
- **Step Two**—Manually isolate the purge steam control valve.
- **Step Three**—Close A and B. Open C. This simulates a complete loss in flow.
- **Step Four**—The rest of the procedure follows the preceding sections.

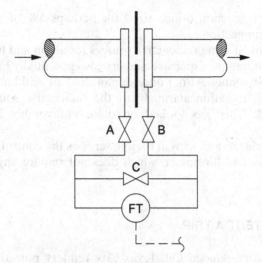

Figure 22-5 *Flow trip or alarm configuration*

In Chapter 19, "Function of a Process Control Engineer," I discuss how to field test that a backup pump would start up automatically, as controlled by a low pump discharge pressure. The control philosophy in this case is to use the more reliable of the two pumps in a backup mode.

OPTICAL SENSORS

"Purple peepers" are optical scanners used to detect flameout of a burner. Burners that go out allow combustible hydrocarbons to accumulate in the firebox. This creates conditions for an explosion. The optical scanner will trip off the fuel gas when the flame goes out. To test the scanner, extract the sensor from the heater and prevent the fuel gas trip to the appropriate burner from shutting off completely. Do not forget to clean the sight glass used by the "purple peeper" sensor.

FLUSHING OF CONNECTIONS

Measurements of flows, pressures, and levels all depend on maintaining small (3/4 to 1-inch) connections open. Alarm and trip connections, as they are backup safety devices, need to be made as reliable as possible. Thus it is good process control design practice to keep the taps flushed. Typically, I use natural gas for vapor connections and diesel oil or steam condensate for liquid connections depending on the service. The flushing medium should best be controlled by a restriction orifice sized for perhaps 5% of the area of the instrument tap connection.

I have never liked using radioactive sources for alarm and trip points. They are quite reliable but are expensive and are also potentially hazardous to the unit operating personnel. Also, I once demolished an acid tank in Texas City with a high-level radiation alarm. I lost the radioactive source during the demolition work. This was looked on quite unfavorably by Amoco Oil management.

For low-flow alarms and trips in fouling services the control engineer could employ an ultrasound flowmeter, which does not require any pressure taps. (See Fig. 21-8)

HOW NOT TO TEST A TRIP

This is a true story, Amoco Oil, Texas City refinery powerhouse—August, 1972.

A 2000 KWH generator was to be shut down to test the over-speed trip on the steam turbine driver. The two operators decided to test the trip as they dropped the electrical generator offline by disconnecting the generator from the electrical grid. In effect, the load on the steam turbine dropped from about 2500 horsepower to maybe 100 horsepower.

The turbine began to run away. The operators successfully tested the over-speed trip. That is, they proved that it was stuck. The generator and the turbine ran faster and faster. Fortunately, so did the two operators. The whole facility self-destructed within moments.

23

Nonlinear Process Responses

Often, changing an independent process variable will result in a nonlinear change in a dependent variable. Sometimes the nonlinear response indicates an equipment malfunction, but sometimes the nonlinear response is quite normal and even desirable, and it always conveys important process information.

Nonlinear responses are different from the creation of such positive feedback loops as:

- Using too much reflux, which makes fractionation worse (Chapter 4)
- Using too little combustion air, which increases energy waste (Chapter 9)
- Effect of CO_2 on global warming (Introduction)

The concept of the positive feedback loop depends on a normally minor process response (such as entrainment from a distillation tray) that gradually becomes the controlling process response (such as jet flood from a distillation tray) as the independent variable is ramped up. The concept of the nonlinear response depends on a single dependent variable responding to changes to a single independent variable. I'll clarify this concept with an example.

- **Forced Condensation**—Figure 23-1 illustrates the correct way to adjust the wash water rate to retard acid attack of the downstream cooler.

Troubleshooting Process Plant Control, by Norman P. Lieberman
Copyright © 2009 John Wiley & Sons, Inc.

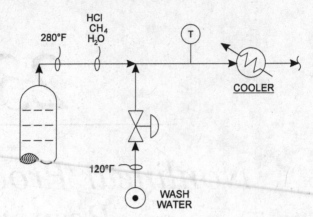

Figure 23-1   *Adjusting wash water to prevent HCl corrosion in exchanger*

Hydrochloric acid (HCl) is in the vapor phase leaving the tower. At 280 °F the steam in the tower overhead is too hot to condense to water. However, when the vapor enters the cooler and first contacts the cold metal tubes, localized condensation of the steam will begin. The HCl is extremely hydroscopic. This means that the first drop of water that condenses will absorb every molecule of HCl acid it contacts. The water will turn into 1 pH acid. The acidic water aggressively attacks and corrodes the carbon steel tubes in the cooler.

To avoid this scenario, wash water is injected into the 280 °F tower overhead vapor line. The wash water cools the overhead vapor. The vapor is cooled not because the water is cold, but because the water evaporates into the vapor phase. The objective is to use enough of this wash water to slightly exceed the "forced condensation dew point temperature." This temperature is defined as the temperature that will cool the vapor to its water dew point. It has the same meaning as when the atmosphere is at its 100% humidity point. That is, the vapor is saturated with moisture at its forced condensation dew point temperature. When saturated vapor enters a cooler, a relatively large volume of water condenses rather quickly on all the tubes. The HCl acid is absorbed in a large volume of water. The resulting acidity of the aqueous phase is now a reasonable 5 or 6 pH instead of a corrosive 1 pH.

- **Temperature Response Curve**—For one crude distillation unit I had calculated that I would have to add enough wash water to reduce the temperature shown in Figure 23-1 from 280 °F down to 220 °F to reach this saturation condition. Monitoring this temperature I slowly increased the wash water flow. I've plotted the results in Figure 23-2. The cooler inlet temperature dropped steadily as I opened the wash water valve. But suddenly, at 240 °F the linear response of the temperature to an increase in

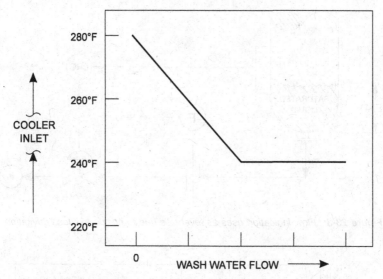

Figure 23-2 *Finding the forced condensation dew point temperature*

the wash water rate stopped. Increasing the water flow further made little difference. The cooler inlet temperature was stuck at 240 °F.

I could not reach my 220 °F target cooler inlet temperature. The reason that the temperature would not decline below 240 °F was that the vapor was already saturated with water at 240 °F. The incremental flow of wash water would not evaporate as the vapor cooling was due to evaporation of water. No further significant reduction of the vapor temperature could be achieved.

In summary, the actual forced condensation dew point temperature was the observed 240 °F, as indicated by the nonlinear response of the temperature to the water flow rate. My 220 °F calculation of saturated dew point temperature was based on flawed data. The stripping steam rate used at the crude unit was not correct.

EFFECT OF NOZZLE EXIT LOSS ON FLOW INDICATION

I observed this particular problem on the Island of Aruba at the former Exxon Lago Refinery. This plant had a dozen hydro-desulfurizers, all of which suffered from the same design error. The problem is shown in Figure 23-3. When liquid drains through the bottom outlet nozzle, the liquid velocity increases from zero to "V," the nozzle exit velocity, in feet per second. The energy to accelerate the liquid comes from the liquid head "H" in inches, according to the relationship $H = 0.34 \times V^2$

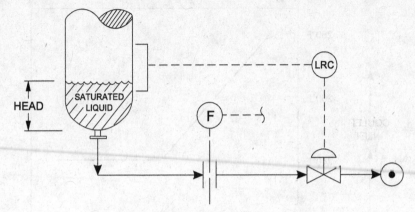

Figure 23-3   *Flow indication rises as level falls due to nozzle exit loss cavitation*

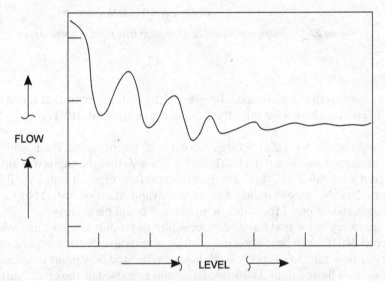

Figure 23-4   *Effect of flashing liquid entering an orifice plate*

If the available head shown in Figure 23-3 is less than the head loss due to acceleration, then the liquid leaving the nozzle will fall below its saturated liquid pressure. The liquid will begin to vaporize.

The bubbles of vapor generated will increase the volume of fluid passing through the orifice flowmeter. The metered flow will rise in the erratic manner shown in Figure 23-4. This indicates to the panel board operator that her flow is going up, when in reality it is probably going down. The lower the level, the greater the indicated flow observed on the panel. Of course, the extra flow is not real. It is just bubbles of vapor.

Placing the flow indicator downstream of the control valve will make the problem worse. The control valve would create more vaporization, which would choke the fluid flowing through the orifice plate.

In Aruba, this was a serious problem. Not because of the nonlinear response of the flow indication to changes in level. The problem was that as the desulfurizer charge rate increased, so did the nozzle exit velocity and the head loss through the nozzle. The flashing liquid caused a large delta P through the flow orifice plate. This restriction then resulted in a high liquid level in the upstream vessel, triggering the vessel's high level alarm to sound.

I fixed this problem by asking the console operator why she needed to know the flow from the bottom of this vessel. Since the flow was erratic and responded in a nonlinear manner to changes in the vessel's level, the operators never referred to the flow. I had the orifice plate removed, and the problem with the vessel's high liquid level bottlenecking the desulfurizer feed rate was eliminated.

NONLINEAR LIQUID LEVEL INDICATIONS

The bottom of distillation towers and vapor liquid separators contains a layer of foam floating on top of a layer of clear, settled liquid. This is shown in Figure 23-5. Let's assume that the layer of foam is several feet above the top tap of the level-trol. Further, let's assume that the layer of foam gradually reduces in density between the foam-liquid interface and the top of the foam layer.

I've actually made such measurements at the Marathon Oil Refinery in Robinson, Illinois. I measured 40 feet of foam in a crude preflash drum. The measurement was carried out with neutron backscatter radiation techniques. The hot crude oil density was 50lbs/ft^3. The foam density varied linearly from 10lbs/ft^3 (top) to 40lbs/ft^3 at the foam-liquid interface.

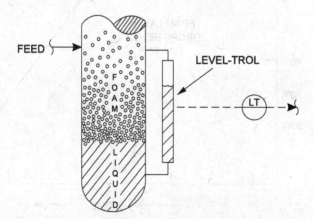

Figure 23-5 *Foam and liquid in vessel, but only stagnant liquid in level-trol*

The light foam was above the vapor-liquid feed inlet nozzle. The force of the vapor entering the vessel entrained the black foam into the top vapor product. This had turned the condensed overhead naphtha black with the entrained crude oil.

The indicated level on the panel was only 70%. That's because this indicated level was based on the height of liquid in the level-trol shown in Figure 23-5. The density of liquid in the level-trol was greater than the density of foam in the vessel. The height of liquid in the level-trol is based on the density of foam between the two level-trol connections. Thus, as long as foam is above the top tap of the level-trol, the indicated level on the panel screen represents not the level in the vessel but the density of foam in the vessel.

Observe a glass of beer. The lighter foam floats on top of the heavier foam. We have the same sort of foam distribution in a vessel. As the panel operator increases the bottom's flow rate, lighter, less dense foam appears between the level-trol connections. The indicated level drops because of a lower foam density and a lower delta P between the level-trol taps. Because the foam density decreases slowly, the indicated level also drops slowly.

However, as shown in Figure 23-6, at some point the indicated level displayed on the panel falls much more rapidly. Using the radiation level scan, I could see what had happened. The actual foam level in the vessel had dropped below the top tap of the level-trol. As shown in Figure 23-6, this occurred at an indicated liquid level of 60%. The break point in the curve that occurs at 60% really indicates that the foam level in the vessel has fallen just below the top tap on the level-trol.

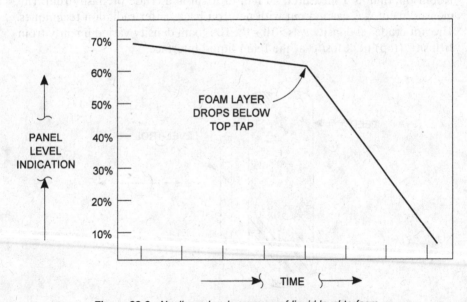

Figure 23-6 *Nonlinear level response of liquid level to foam*

The foam level was now also below the vessel's feed inlet nozzle. A sample taken an hour later of the overhead naphtha product was clear, clean, and free of entrained black crude oil.

The confusing aspect of this problem is that the indicated level drops slowly in a normal manner as the panel operator increases the bottom's flow rate. This makes the operator think that everything is normal, and that the indicated liquid level is correct. The sudden nonlinear response of the indicated level shown in Figure 23-6 is wrongly interpreted by the panel board console operator as an instrument malfunction. But in reality it is an indication of the foam layer dropping below the upper level-trol tap on the vessel.

CENTRIFUGAL PUMP DISCHARGE PRESSURE

The Process Control Engineer typically assumes:

- A small decrease in pump flow due to closing the discharge control valve will result in a small increase in pump discharge pressure.
- A small decrease in pump suction head pressure due to a decrease in liquid level will result in a small decrease in the pump discharge pressure.

For many larger high-head pumps working at low suction pressures, neither assumption is always true. For instance, at flows well below design, closing the pump discharge control valve DECREASES the pump discharge pressure, certainly a nonlinear and unexpected result that causes the design of control loops to be difficult at best.

Also, I have seen many large centrifugal motor-driven pumps operating close to their minimum required net positive suction head. A reduction of 1 foot of liquid level then causes the pump to lose 50% of its discharge head but still retain a stable flow and a stable discharge pressure, again, an extreme nonlinear response to a small change in liquid level. The services where I've observed these difficulties are for pumps running at 1–10 psig suction pressure, a flow of 500–2000 GPM, and discharge pressures of 150–250 psig.

These and many other nonlinear parameter responses to process changes are usually considered instrument malfunctions. Often, however, these nonlinear responses are important indications of critical points in operating parameters. Thus they can be used to advantage by both the Process Control Engineer and panel board operators to troubleshoot complex process plant control problems.

About My Seminars

I've been working as a process engineer for 44 years. Fortunately for me, technology, as it affects the process industry, has not changed in the interim. The technology that affects our work is:

- AC motor—1890
- Centrifugal compressor—1930
- Centrifugal pumps—1900
- Thermodynamics—1880
- Vapor–liquid equilibrium—1910
- Continuous distillation—1840
- Internal combustion engine—1880

All the catalytic technologies we use (alkylation, hydrotreating, reforming, isomerization, fluid catalytic cracking) were old and well established when I came to work for Amoco Oil in Whiting, Indiana in 1965. Thus my seminar represents today's technology even though I've learned almost nothing new in 44 years, except for one idea.

WHAT'S NEW

Four billion years ago, before I was born, the content of the atmosphere was:

- 79% N_2
- 21% CO_2

Plant life converted the CO_2 to O_2 and hydrocarbons. The hydrocarbons were buried as:

- Methane hydrates (buried off-shore on continental shelves)
- Coal
- Shale oil
- Tar sands
- Oil
- Gas
- Peat

The above list is in descending order of abundance. Currently we are extracting these deposits at a rate of 210,000 FOE B/D (fuel oil equivalent barrels per day). At this rate, we will have reconverted all of the available O_2 back to CO_2 in 35,000 years. Of course, that's nonsense. CO_2 becomes fatal when its concentration exceeds 3%, not 21%. For life like us, normal respiration would be impeded at CO_2 levels above 1%. So, we have one-twentieth of 35,000 years, or 50 human generations, to correct the problem.

SYNERGISM

Of course, this too is nonsense. The largest amount of hydrocarbons are tied up not in fossil fuels (oil, gas, coal), but in hydrates. These hydrates (a combination of light hydrocarbons and water) that freeze at temperatures between 40 °F and 60 °F can be released because of global warming. Unfortunately, 1 mole of methane equals 23 moles of CO_2 as a greenhouse gas. Without exception, all of the extinctions in history have been accompanied by a surge in greenhouse gases, usually triggered by meteor impacts and volcanic activity.

WHO CAUSED THE PROBLEM?

People like you and me. To be specific:

- Newcomb, Watt—Steam engine and barometric condenser
- Tesla—Alternating current 3-phase motor

- Tesla—Radio, Remote Control, Florescent lighting
- Otto, Diesel—Internal combustion engine
- Whittle—Jet engine, gas turbine
- Tesla, Edison—Electric lighting

Like us, these guys were technical nerds. I used to teach in my seminars that my primary interests were:

- Food
- Sex
- Money

This was a lie. My only interest has always been, right from childhood, technology—just like you. Technology for its own sake, regardless of its consequences. It's an expression of our instinctive desire to dominate nature, regardless of its consequences.

CONSERVATION IDEAS IN THE SEMINAR

I'm trying to atone for my sin. The source of sin is acting without regard for the consequences of our actions. That is why I've written this book. I believe that if the control techniques I've described in this text and the process design and operating concepts I explore in my seminar are used, hydrocarbon waste will be reduced. The central concept of my troubleshooting seminar is a search for ideas once well known, but long since forgotten. This book is an expression of what I have discovered in that search.

Norm Lieberman
New Orleans, Louisiana—May 28th, 2008.
1-504-887-7714 (phone)
norm@lieberman-eng.com (E-mail)

Further Readings on Troubleshooting Process Controls

Instrumenting a plant to run smoothly, *Chemical Engineering*, Sept. 12, 1977, N. Lieberman.

Distillation Control for Productivity and Energy Conservation, McGraw-Hill, 1984, Shinskey.

Process Control Systems, McGraw-Hill, 1996, Shinskey (highly recommended basic text).

Principals and Practice of Automatic Process Control, John Wiley & Sons, 2005, Carlos Smith.

Working Guide to Process Equipment—3rd Edition, McGraw-Hill, 2008, N. Lieberman and E. Lieberman.

Distillation Operation, McGraw-Hill, 1990, H. Kister (many excellent case histories).

Distillation Simulation for Design and Control, John Wiley & Sons, 2006, W. L. Luyben.

Process Industrial Instruments and Control Handbook, McGraw-Hill, 1992, McMillan & Cosidine.

Troubleshooting Process Operations—4th Edition, Penn Well Publications, Tulsa, 2009, N. Lieberman.

DISTILLATION

	Minutes	Price
Distillation Fundamentals	42	$450
Jet Flood & Dry Tray Pressure Drop	30	$320
Incipient Flood—A Basic Operating Concept	40	$250
Identifying Flooding Due to Foaming	24	$350
Tray Dumping & Weeping	32	$400
Tower Pressure Drop Evaluation	31	$400
Plugging & Fouling of Trays	41	$400
On-Stream Cleaning of Trays	32	$250
Composition Induced Flooding	25	$250
Packed Tower Fundamentals	31	$400
Structured Packing Pitfalls	35	$400
Optimizing Pumparound Rates	28	$350
Coker, FCU & Crude Fractionators	41	$480
Inspecting Tower Internals	27	$250
Effect of sub-Cooled Reflux	25	$250
Excess Feed Preheat	25	$250
Troubleshooting Steam Strippers	40	$400

HEAT EXCHANGERS & REBOILERS

	Minutes	Price
Heat transfer Fundamentals	43	$320
Once-Through Thermosyphon Reboilers	23	$200
Circulating thermosyphon Reboilers	25	$250
Kettle Reboilers	29	$250
Forced Circulation Reboilers	25	$250
Foam Formation in Reboilers	20	$300
Steam & Condensate Flow	42	$300
Water Hammer—Causes & Cures	18	$220
Shell & Tube Exchanger Fouling	33	$280

FIRED HEATERS

	Minutes	Price
Fired Heater Fundamentals	31	$400
Draft Balancing Heaters	30	$300
Optimizing Excess Air in Heaters	34	$400
Preventing Heater Tube Coking	34	$400
Air Preheaters	35	$350

CONDENSERS

	Minutes	Price
Fundamentals of Condensation	33	$300
Back-Flushing & Acid Cleaning Condensers	20	$220
Pressure Control of Columns	30	$300
Sub-Cooling Robs Condensers Capacity	30	$250
Surface Condensers	39	$400

VACUUM TOWERS

	Minutes	Price
Improving Vacuum Tower Operation	40	$450
Packing & Grids in Vacuum Towers	35	$350
Steam Jets	29	$300
Vacuum Tower Overhead Systems	41	$350

PUMPS

	Minutes	Price
Fundamentals of Centrifugal Pumps	32	$250
Optimizing Impeller Size	25	$200
Cavitation & NPSH	31	$450
Causes of Seal & Bearing Failures	31	$400
Erratic Pump Discharge Pressure	30	$300

COMPRESSION

	Minutes	Price
Fundamentals of Gas Compression	35	$350
Surge in Centrifugal compressors	25	$400
Variable Speed Centrifugal	20	$250
Compressor Rotor Fouling	30	$300
Steam Turbines	24	$300
Reciprocating Compressor Efficiency	24	$450
Compressor Motor Over-Amping	27	$250

TREATING

	Minutes	Price
Amine Regeneration & Scrubbing	42	$400
MEA Degradation	27	$300
Sulfur Plant Start-Up	20	$250
Sulfur Plant Pressure Drop	40	$400
Jet Fuel treating	33	$300

REFINERY PROCESSES

	Minutes	Price
Troubleshooting Sulfuric Acid Alky	35	$450
Crude Pre-Flash Tower	30	$350
Crude Tower Overhead Corrosion	30	$300
HF Alky Pressure Problems	25	$250
Troubleshooting Delayed Cokers— Part I	59	$500
Troubleshooting Delayed Cokers— Part II	62	$500

SAFETY

	Minutes	Price
Field Testing Alarms & Trips	33	$280
Failure of an Amine C_3-C_4 Scrubber	36	$380
Detonation at an FCU	30	$330
Boiling Water Causes Coker Fatality	35	$380
Sulfur Plant Safety Hazards	30	$300
Sources of Auto-Ignition in Crude Units	20	$250

ADD $10.00 PER TAPE FOR SHIPPING OR ADD $35.00 PER TAPE FOR AIR FREIGHT

AIR FREIGHT RATE REDUCED FOR BULK ORDERS

— Prices in US dollars for US ½" VHS NTSC.
— Each tape may be purchased separately.
— A set of free textbooks is provided with each Order of ten or more tapes.
— For 3/4" or format other than ½" US VHS NTSC Add 10%.

PROCESS CHEMICALS, INC.
PMB 267
5000 W. Esplanade
Metairie, Louisiana 70006, USA
(504) 887-7714
Fax—(504) 456-1835
http://www.lieberman-eng.com/

Index

Troubleshooting Process Plant Control, by Norman P. Lieberman
Copyright © 2009 John Wiley & Sons, Inc.

Printed in the United States
By Bookmasters